EARLY
AMERICANS

The Beginnings of the American People
I. Vexed and Troubled Englishmen, 1590–1642
II. No Peace Beyond the Line:
The English in the Caribbean, 1624–1690
(with Roberta Bridenbaugh)

Jamestown, 1544–1699

*The Spirit of '76: The Growth of American Patriotism
before Independence, 1607–1776*

*Fat Mutton and Liberty of Conscience:
Society in Rhode Island, 1636–1690*

Silas Downer, Forgotten Patriot: His Life and Writings

*Cities in the Wilderness: The First Century
of Urban Life in America, 1625–1742*

Cities in Revolt: Urban Life in America, 1743–1776

Peter Harrison: First American Architect

The Colonial Craftsman

Myths and Realities: Societies of the Colonial South

*Mitre and Sceptre: Transatlantic Faiths,
Ideas, Personalities, and Politics, 1689–1776*

Seat of Empire: The Political Role of Williamsburg

Rebels and Gentlemen: Philadelphia in the Age of Franklin
(with Jessica Bridenbaugh)

*Gentleman's Progress:
The Itinerarium of Dr. Alexander Hamilton, 1744*
(edited, with an introduction)

*The Transactions of the American Philosophical Society, &c.
Published in the American Magazine During 1769*
(facsimile reprint, with an introduction)

*A Tour Through the North Provinces of America . . . 1774
and 1775*, by Patrick M'Robert (edited, with an introduction)

The Partisan Leader: A Tale of the Future,
by Nathaniel Beverley Tucker (edited, with an introduction)

EARLY AMERICANS

Carl Bridenbaugh

New York   Oxford
OXFORD UNIVERSITY PRESS
1981

Copyright © 1981 by Carl Bridenbaugh

Library of Congress Cataloging in Publication Data

Bridenbaugh, Carl.
Early Americans.

1. United States—History—Colonial period, ca.
1600–1775—Addresses, essays, lectures. I. Title.
E188.5.B74   973.2   80-26313
ISBN 0-19-502788-4

Printed in the United States of America

For
EDMUND
and
HELEN MORGAN

Preface

The following essays were written over nearly half a century. When they were brought together for this volume, I was surprised to find that, in one way or another, they all had to do with the comings and goings of colonial "chaps" which, the English insist, is what history is all about. Hence the title of this book: *Early Americans*. With two exceptions, the essays are reprinted as they appeared in several learned publications or magazines where, as Carl Becker once wrote appropriately, they had been so decently interred. The first chapter on Opechancanough and the fifth about Tom Bell have not been published before.

For permission always graciously accorded to republish articles that appeared first in their journals, I am grateful to Marcus A. McCorrison of the American Antiquarian Society, Nicholas B. Wainwright of the Historical Society of Pennsylvania, Louis L. Tucker of the Massachusetts Historical Society, and the editors of the *William and Mary Quarterly*.

Providence   C.B.
March 1, 1980

Contents

List of Plates

Captain Smith worked with Vaughan, who thereby was more accurate in his representations than most of the Europeans. Plate 1 was published in Germany and Plate 2 in England in 1624 and 1675. They reflect the Virginia scene as viewed and imagined by white men and, therefore, are not accurate likenesses. But they are the only visual records of actual events that we possess. Above all else, they testify to the fact that white men from overseas always looked at the native Americans through European spectacles. (Courtesy of the John Carter Brown Library)

EARLY
AMERICANS

Introduction

The writer's primary purpose—or hope—in reprinting the pieces that follow is to entertain thoughtful, general readers with a bent toward colonial history. Here are nine vignettes involving people individually or in groups in the American colonies of England over a period of more than two centuries. These essays were originally written to stress or describe in detail particular, underemphasized or unknown features in the growth of a kind of agricultural-commercial society that was, in many ways, radically different from the societies of the British Isles or Western Europe yet at the same time strikingly resembled them.

"Americans are always moving on," Stephen Vincent Benét has told us, and each of these essays bears out his pronouncement. In the seventeenth century (Chapters II–IV), we perceive that the majority of the colonists were immigrants, foreigners who had embarked on their peregrinations from across the great ocean. After 1700 this was less true, but the geographical stage on which the colonial travelers roamed was vast—from the Province of Maine overland to Georgia and by sea from the ports on the Atlantic coast to the British Caribbees.

The amount of intercolonial travel for business or pleasure is startling; small at first, it mounted progressively as the decades passed. The roads, though crudely built were passable, and the people who traveled on them were as motley a crew as Chaucer's band of Canterbury pilgrims. The relative safety in which women, as well as men, could make their way through sparsely settled areas is amazing. As early as 1704 Madam Sarah Kemble Knight rode, unescorted for the most part, from Boston to New York and back without incident. In 1744, Dr. Alexander Hamilton and his body slave Dromo journeyed leisurely to and from Annapolis and Kittery in great safety. "The Father of His Country" seemed almost to have spent more time in the saddle during the 'fifties and 'sixties than out of it. The growing complexity of life in America and the surprising variety of colonists encountered on the move may be discovered from the journals of these and other travelers. A mounting internal migration of individuals was under way, and by 1760 there was taking place a remarkable mingling of the inhabitants of the thirteen colonies that was breaking down some of the provincialism and creating a human type that observers could recognize as American, not English any more.

The first four chapters of *Early Americans* deal with immigrants, most of them English and Irish, and their encounters with the native Americans, their initial struggles in building shelters and houses, and the founding of little settlements—in short, the forming of societies of the English type in the New World setting. Chapters V-IX treat important aspects of colonial life in the eighteenth century and the astonishing rapidity with which European, rather than just English, habits and new modes of living were imported and given a colonial twist. This was true of the *bad*, as well as the *good*, that the Old World had to offer. And one can begin to observe and understand the accelerated growth and rapid maturing of colonial America.

I

O-pe-chan-can-ough:
A Native American Patriot*

History always affects present-day issues and events to some degree. The *Indian problem,* specifically the issues arising out of the claims to the Narragansett, Mashpee, and Passamaquoddy lands, is very much with us right now; and if we look backward to the very first accounts of the "tawny" native Americans with the white English invaders, we can see at once that the *Indian problem* arose long before the settlement of Jamestown in Virginia in 1607. It became, and remains, part of the national heritage.

The historian's great difficulty is that he has only the white men's accounts for use in any reconstruction of what happened in seventeenth-century Virginia, and these are fragmentary, biased, often inaccurate, and always inadequate. Very little new information has become available since 1900. Nevertheless, despite the deficiencies in the record, he must take a chance with such sources as he has. By a fresh reading of the known facts and a rigorous use of a disciplined historical imagination, one can construct a *plausible hypothesis* that will suggest what the Indian point of view was as well as that of the white man.[1]

The colonists who came to Jamestown in 1607 with the professed purpose of propagating the Christian religion and bringing

* Expanded from the Penrose Memorial Lecture delivered to the American Philosophical Society at Philadelphia, April 17, 1980.

the "savages" of Virginia to human civility had no idea that the natives had known white men over a long stretch of time and that they both mistrusted and feared them. Nor did the Englishmen ever really grasp the nature of the hazard inherent in the confrontation of the two races.

Within five years of the landing of the English, the secretary of the colony of Virginia, William Strachey, had made himself the best authority on the "naturals." He had found out, among other things, that members of the Algonkian tribes dwelling near Jamestown were known to speak frequently about CERTAIN PROPHECIES that, kept alive and circulated by their priests, profoundly affected Powhatan and his "kings" in their dealings with the English.* The first of these told that "from the Chesapeack Bay a Nation should arise which should dissolve and give end to his Empire." Just a short time before Captain Christopher Newport's shore party was attacked near Cape Henry on April 26, 1607, Powhatan's braves had wiped out the members of the Chesapeake Nation living in this region and with them, as Strachey learned, a number of white men from Raleigh's "Lost Colony" of Roanoke who had lived among the Chesapeakes for about two decades.[2]

A second prediction was that *twice* the Indians would defeat and drive away the strangers who would invade their territories and "labor to settle a plantation" in their midst, but the *third* time they themselves would be decisively conquered. Fear for their safety and future existence had impelled Powhatan and his chiefs to maintain an elaborate watch over every coming and going of the shiploads of white men landing upon their coasts, which were supplemented each year by "fresh troops." Strachey reported that "strange whispers (indeed) and secret at this hour run among these people and possess them with amazement" and apprehension. Assailed by doubts as to the proper course to pursue, Powhatan, "the great Tyrant," seems to have lost confidence in his wonted procedures, and the secretary concluded that, with "divine power" on their side, the English could effect mighty changes—"accidents," he called them—in Virginia.[3]

Powhatan was the native place of Wa-hun-son-a-cock, who, when he became the principal chief of many Algonkian tribes in Tidewater Virginia, took the name of the town. Succeeding chiefs were known as *the Powhatan*. Powhatan was also the name of the tribe. In this essay *Powhatan* will be used in all three senses, the particular meaning being indicated by the context.

Although William Strachey obviously regarded these prophecies as primitive myths, today we can look upon them as *oral history*, a means by which an illiterate but intelligent primitive people keeps the past alive to guide it in future action. From the evidence we have, it is clear that certain historical incidents justified the Indians' anxiety; and, moreover, the first prophecy turned out to be true. In addition to the legends and the suspicion that the English represented the *third invasion* by strangers, there was, behind it all, an almost incredible and fascinating story.

Ever since the year 1607, Powhatan, the despotic ruler of many tribes living about Jamestown, has been deemed by both the public and historians to be the foremost native leader in Virginia during the seventeenth century. Today we may consider not only whether an elder brother, *Opechancanough*, did not far surpass him in talents and capacity for leadership but also if Opechancanough does not deserve to rank high among the most famous American Indians—with Massasoit, King Philip, Pontiac, Logan, Joseph Brant, Sitting Bull, Crazy Horse, and Geronimo.

This brother was born in or about 1544, for in 1644 it was widely believed that he was more than a hundred years old. The longevity of the American Indians never ceased to amaze the English settlers, and this particular individual was far from unique.[4] Early in 1561 when he was sixteen or seventeen and already, by European standards, a very tall young man, two small Spanish ships commanded by the famous mariner Pedro Menéndez de Avilés discovered and entered Bahía de Santa María (Chesapeake Bay) on their way back to Spain from Havana. The "Admiral" was under orders from Luis de Velasco, Viceroy of New Spain, "to discover what ports there are at the Punta de Santa Elena [Parris Island, S.C.], and the coast for eighty or one hundred leagues from there on towards Los Baccalaos [Newfoundland]." As soon as the natives saw the vessels drop their anchors, they paddled out in their canoes and boarded Menéndez's flagship, where he generously fed them and gave them clothing. "Among these Indians came a chief who brought his son, who . . . was of fine presence and bearing." (Plates 1, 2) Immediately drawn to the youth, the Spaniard sought permission to take him to Europe "that the King of Spain, his lord, might see him . . . *He gave his pledged word to return him* with much wealth and many garments." The chief granted this, and his Excellency took him to Castile, to the court of

King Philip II late in the spring or early in the summer of 1561.[5]

Pedro Menéndez de Avilés, himself a nobleman, introduced the young American as a *"cacique* or important lord from . . . Florida."* Astonished by his imposing size, fine physique, and high intelligence, His Most Catholic Majesty and his grandees showed themselves to be "very pleased with him." Urged, no doubt, by the zealous, persuasive Menéndez, who had soon to depart for the West Indies, the pious monarch not only ordered an allowance for supplying the chieftain with rich clothes befitting his noble condition, but he also seems to have placed him with the friars of St. Dominic at Seville to be taught the Spanish language and instructed in the elements of Christianity. The king and courtier hoped that ultimately he could rejoin his fellow tribesmen, as promised, and help to prepare them to receive the Holy Faith. Herein lay great possibilities of which the quick-witted and perceptive "savage" would make much.[6]

In 1561, or the following year, Menéndez sailed as captain-general of the galleon fleet for Mexico, and, in the meantime, the neophyte (still nameless in the records) made real progress with the Castilian tongue and his religious studies. All writers testified to his superior intelligence and that he was both "wily" and "crafty." He also proved to be unusually observant, alertly accumulating a body of knowledge about the white men's vast numbers, modes of living and social organization, their wealth, technology, military strength, and methods of fighting. And, portentously, he appears to have understood much of the meaning and relationships of what he had learned. This was especially true about the forehandedness of the Spaniards and their capacity for long-range planning of great activities—qualities conspicuously lacking among the Indians of North America.[7]

Under Dominican tutelage the pagan soon became a Christian. There is a possibility that his first meeting with the Viceroy of New Spain, Don Luis de Velasco, took place at the Spanish court and not in Mexico (as most authorities surmise). In any event, he made so great and lasting an impression upon the viceroy that the latter undertook to act as godfather to the American and gave him his own name. "Thus the Indian son of a petty chief of Florida was called Don Luis," the Jesuit Juan Rogel observed with characteristic European condescension.[8]

"Admiral" Menéndez returned from the New World in 1563,

and when he sailed westward again he took Don Luis along. Both
the king and the mariner wanted the pledge to return the Indian
to Bahía de Santa María redeemed, and Philip II had directed that
he be sent homeward. Likewise, Don Luis was, understandably,
homesick and eager to see his kinsmen once more. It is not fanciful
to picture the two men on board the flagship during the long
westward passage discussing the Spanish grandee's plans for a
Christian settlement in Ajacán (as Don Luis named his country),
and the Indian's calculated emphasis upon the preparatory role
that he could play. Once in Mexico, Menéndez proposed to Governor Velasco that Don Luis be sent home promptly, but the Archbishop of New Spain, fearing that the convert might revert to
paganism, opposed such a move. Thus outmaneuvered, Menéndez
left the youth with the Dominicans for the time being.[9]

After this setback, the "Admiral" was busily occupied with
ideas and plans for a Florida "Empire." Prominent among these
were the searching out a passage through the continent of North
America, which he had learned of by 1561, beginning somewhere
in Chesapeake Bay, and the founding of mission colonies among
the native populations of Florida including the future Virginia. The
colonies were to be fortified, and Menéndez wrote to Philip II on
October 15, 1565, that Bahía de Santa María was to be the key to
all defense since beyond it lay Newfoundland and the menacing
French. With regard to this second project, he believed that the
Indians of the land bordering the great bay gave "evidence of good
intelligence" and were "not as rustic or savage as the others. Thus
they can boast of established customs. In conformity with their
idea of justice, they punish liars, although in all reality all Indians
are liars, and they abominate thieves." Moreover he had very
much on his conscience the promise of 1561 to restore Don Luis to
his people, and the Indian, we know, had from time to time reminded him subtly of this obligation by pointing out the great help
he and his "three brothers" could be in the holy work of conversion. For approximately three years longer, the native remained in
Mexico with the Dominicans, during which time he had an excellent opportunity to acquire some understanding of the way the
Mexican Indians were treated, as well as more about Spanish colonial life, neither of which would have appeared encouraging to
him.[10]

King Philip signed a contract with Pedro Menéndez de Avilés

on March 20, 1565, to add royal funds to the nobleman's own large fortune and granting him permission to go out as *Adelantado* or conqueror of Florida, which was still part of the Viceroyalty of New Spain. Two days later Menéndez shrewdly forestalled possible archepiscopal opposition by procuring a royal order to the President and *Audencia* of New Spain "upon demand" to remand Don Luis to his custody. On June 28 he set out from Spain on his new mission. Upon reaching Havana, the intensely zealous Roman Catholic *conquistador* quickly moved against the Huguenots at Fort Caroline and, during September and October, cruelly and ruthlessly killed or captured all of the French in Florida. After that he was ready to implement his grand design. From Havana on December 25, 1565, he dispatched Philip II's order by his lieutenant, Dr. Solís de Merás, to Mexico when the latter went there to solicit material and religious aid for Florida. Bowing to the supreme authority, both the Archbishop and Governor Velasco saw to it that Don Luis was sent to San Mateo in the company of two Dominican friars.[11]

The *Adelantado* himself went to San Mateo (the former Fort Caroline, then rebuilding) and from there on August 1, 1566, issued instructions for a colonizing missionary expedition to Bahía de Santa María under the leadership of Fray Pablo de San Pedro and another Dominican, both from Mexico. In the orders Menéndez stressed "that it is in the service of God our Lord and of his Majesty that I send Don Luis, the Indian, to his country, which according to him is between the 36th and 39th degrees [north latitude] along the shore, and all the people of that territory are his friends and the vassals of his three brothers." Then, proceeding to Santa Elena, the governor directed Captain Pedro de Coronas with thirty soldiers and the two friars to sail at once to colonize and set up a mission at Ajacán, the land around Bahía de Santa María. In his life of Menéndez (written the next year), Bartolomé Barrientos wrote that "with his party went a brother of the *cacique* of that area. This man the *Adelantado* had taken from there six years before. He was very cultivated, showed good understanding, and was a good Christian. His name was Don Luis de Velasco, and he was sent to help the conversion of the natives." By taking such action, the grandee from Asturias would naturally have felt that he was both doing the Lord's bidding and honorably fulfilling a promise made years before.[12]

Guided by Don Luis, the expedition missed the Chesapeake capes and instead anchored on August 14, 1566, at 37° 30' north, possibly in Chincoteague Bay. The youthful native's failure is easily explained if one recalls that he had left his village at the age of sixteen or seventeen and seen the entrance to the bay only once— six years earlier—when he sailed outward with Menéndez. After being driven southward by a storm to latitude 36°, exploring North Carolina waters, and landing somewhere on the Carolina Outer Banks, Domingo Fernandez, the pilot, sailed northward again to 37° 30' without Don Luis being able to recognize his land. When a wind storm lasting four days blew *La Trinidad* out to sea, Fernandez advised setting a course for Spain rather than attempting a return in hurricane season to San Mateo or the Windward Islands. Captain Coronas and the two friars concurred willingly with his recommendation but apparently did not consult Don Luis. Not to reach his home when he had come so close to it and after such a long absence must have been a bitter disappointment, and his being ignored by the white men, including two of the "religious," must have been a grievance a proud man would never forget.[13]

La Trinidad stood into Cadiz harbor on October 23, 1566, and Don Luis entered upon a second Spanish sojourn. The two Dominicans took him to Seville on October 30 on their way to the court at Valladolid. Although there is no explicit record, it may have been King Philip himself who turned Don Luis over to the Jesuits for further education, for that religious order was now considered the most active in the colonial field as well as in teaching. Further, Pedro Menéndez de Avilés may have had a part in the decision, because he had been courting Jesuit support for his plans since 1565. Father Geronimo de Oré, a Franciscan historian, tells us that about 1568, "the *cacique* Don Luis was living in the [Jesuit] house at Seville, *advancing in the Spanish language, both in reading and writing, together with other branches of knowledge* which they taught him." Sometime before the end of this second period in Spain, which lasted more than three years, Bartolomé Martinez related of Don Luis: "Being intelligent (as the Indians of those [American] provinces are, if they mingle from youth with the Christians) he was made ready and they gave him the holy sacraments of the altar and Confirmation."[14]

During the autumn of 1567 while Menéndez was once more in Spain, the twenty-three-year-old American Indian wished more than ever to return to his land of Ajacán. He broached to his Jesuit mentors a long-cherished "plan and determination, as he then said, of converting his parents, relatives, and countrymen to the faith of Jesus Christ, and baptizing them and making them Christians as he was." When he heard that some "Religious had gone for the conversion of the natives of Florida, he told Father Rector and others he [Don Luis] would venture to take some priests to his country, and that with the help of God and his own industry, the Indians of that land would be converted to the Faith." The Fathers liked the proposal, and the General of the Society of Jesus, Pope Pius V, and King Philip all consented to such an expedition. As for Menéndez, naturally he was delighted—and relieved no doubt. He knew personally the land round Bahía de Santa María and he offered to transport the Jesuit missionaries and Don Luis thither. Once again he saw an opportunity to carry out his promise to the Indian's father and further the Lord's work.[15]

The *Adelantado,* again placed in command of the galleon fleet, dispatched Father Juan Baptista de Segura and a party of Jesuits ahead of him. They reached San Agustin on June 29, 1568, but Menéndez did not arrive at Havana until November. With him was Don Luis, fresh from his fourth and final Atlantic crossing. He had also made at least two long voyages between Havana and Vera Cruz, as well as coasting trips between San Agustin and the entrance to Chesapeake Bay. It is inconceivable that during these long and tedious passages this highly intelligent young man would have failed to become familiar with the way of a ship, the devastating fire-power of marine artillery, and that during the many hours spent in the grand cabin or on the deck with Pedro Menéndez de Avilés he would not have satisfied his curiosity about the instruments of navigation he saw used daily, particularly the fascinating mariner's compass. Such knowledge would serve him admirably in after years.[16]

Father Segura, the Jesuit vice-provincial, had gone from San Agustin to Havana before the arrival there of the galleons, shortly after which the *Adelantado* brought Don Luis to see him. The Indian explained to the priest that he would like to accompany the Jesuits to Ajacán in the capacity of both guide and interpreter. It was said of the missionaries that "practically every one was against it." Nevertheless the priest decided to use Don Luis, because "it was thought that an important person with many relatives, familiar with the land, could help Ours in converting his subjects and friends, as he had promised." Preparations took up considerable time during which Menéndez received a letter on August 18, 1570, from Pope Pius V exhorting him to get on with the work of propagating the faith in Florida. Actually, in August, Vicente González had sailed from Havana for Santa Elena, where the missionaries joined him, and landed "in the land of Don Luis," Ajacán, on September 10.[17]

The Indian had pictured his country truthfully as a land of great plenty, but to their dismay the Jesuits found that for several years the natives had been afflicted by both disease and famine, and were still very short of food. However, they proved very kindly, thinking, as Fathers Quirós and Segura reported on September 12, "that Don Luis had risen from the dead and come down from heaven." Likewise two of his *cacique* brothers received Don Luis warmly; they informed him that his elder brother had died and that "a younger one was ruling." Almost certainly this was the chieftain known to history as Powhatan. He offered at once to relinquish all of his authority. Thereupon Don Luis declined to assume the title and power of the Powhatan, "asserting that he had not returned to his fatherland out of a desire of earthly things but to teach them the way to heaven which lay in instruction in the religion of Christ Our Lord. *The natives heard this with little pleasure,*" the reverend chronicler reported.[18]

In a postscript to the letter quoted above, Father Quirós added: "Don Luis turned out well as was hoped, he is most obedient to the wishes of Father [Segura] and shows deep respect to him, as also to the rest of us here. . . ." But very shortly, probably before October 1, the lusty young American "now in the midst of his own people"—like many a "civilized Indian"—found the freedoms and customs of aboriginal life irresistible and reverted to old ways. Openly breaking with the canon law dictum of "one wife, in perpetuity," he immediately "took to himself as many wives as the Gentiles." In the eyes of his fellow-tribesmen, his conduct was both morally approved and expected of a werowance or "king," but to the pious missionaries, it was clearly the wicked work of the Devil.[19]

According to the "Relation" of Father Oré, Father Juan Baptista de Segura reprimanded the sinner severely in front of everyone at the mission. "Afterwards, in words couched in the spirit of religion and charity, he admonished him, telling him that he should remember that they had come, moved by the promises he [Don Luis] had made in Spain, and under his protection. Moreover, if he gave such a bad example, they would not be able to implant the Gospel, in whose ministers and interpreters cleanness of life is so important. Despite these and other gentle words, which he [Father Segura] and other Religious spoke to him, they could not soften him, but rather they were the occasion of spiritual hardening of the heart."[20]

Publicly upbraided and humiliated and privately dressed-down for his carnal sins by Father Segura, the Indian concluded that he would renounce Christianity and resume his pagan faith and ways. Wishing to withdraw from the sight of him who had so severely reproached him, "Don Luis said that he was going to look for chestnuts and nuts of other varieties, in one of his towns . . . and he would return very soon." Leaving the tiny mission on the York River in October, he "went native" in the village of his brothers on the Pamunkey River. After about four months, when he failed to rejoin the mission, the Jesuits instituted a search: Father Quirós and two others were sent by their leader up the Pamunkey to urge Don Luis to go back with them. Ostensibly agreeable, he promised he would return. He followed them with a number of Indians, but when they caught up with the missionaries on February 4, 1571, they slew them with a "shower of arrows," burned their bodies, and made off with their clothing and bundles. Not satisfied with this vindictive act, the natives proceeded to the mission and, on February 9, murdered Father Segura and the remaining company (Plate 1) with the exception of one boy, Alonso de Olmos, who escaped and survived to tell the tale to members of a Spanish punitive expedition in 1572.[21]

These events signaled the complete and permanent break of Don Luis with the white Christians. Mortified by Father Segura's public censure of him, fully aware of the hostility of his own people to Christianity of the Roman Catholic variety, and desirous of once more enjoying power suitable to his talents, experience, and ambitions, Don Luis went back to live with and work for his own race. Symbolic of this determination to drop the dual role he had been

Don Luis de Velasco Murdering the Jesuits, 1571

playing was his abandoning the name of Don Luis de Velasco and assuming a new one—significant in view of his past adventures and arcane knowledge of the strange white people—OPECHANCANOUGH, Algonkian for "He whose soul is white."[22]

Members of a relief expedition headed by Vicente González sent to the defunct mission in the spring of 1571 discovered the ghastly truth about the Jesuit fathers when they saw the Indians on shore walking around in cassocks. A landing party captured two chiefs and then González sailed away. From one of their captives, the Spaniards learned some details of the massacre and of the existence of the boy Alonso. The next year, in September, Pedro Menéndez de Avilés himself arrived with a punitive party, and they seized thirteen important natives, including a chief, Opechancanough's uncle, and slaughtered more than twenty. The Spanish also rescued Alonso, the sole survivor of the mission. He had not witnessed the killings but had learned most of the details while under the protection of the chief (probably Powhatan) when Don Luis was trying to eliminate all white men in the region. Don Luis escaped into the woods and consequently his uncle and the hostages were all adjudged guilty and hanged.[23]

It is not stretching the imagination or pressing the facts too far to assume that after the bloody work of 1571, Opechancanough provided his brother with all sorts of details about the white Spaniards that both troubled and frightened Powhatan. And one can easily believe that such conversations, which began some thirty-odd years before the arrival of the English on the James River, help to explain the legends about Opechancanough having come from the southwest or West Indies and the prophecies of the destruction of the tribes by foreign invaders. During these same years, 1571–1607, Powhatan's empire was being built up and strengthened by conquest, and his capable brother must have played a major part in its formation, for in the words of the Europeans, he was the heir apparent. The policy worked out by these chieftains for dealing with the white strangers was adopted and perpetuated by the native priests in their "prophecies." At the time of the killing of the Jesuit missionaries—Opechancanough's *first* massacre—both he and Powhatan were fully apprised of the foreign menace that augured the end of their power and possibly of their tribal existence. As for the returned native, he must have regretted many times that he had rejected the great office of the Powhatan, leaving it in the hands of his naïve and often indecisive brother.[24]

Unfortunately the natives left behind no record of their experiences with and feelings about the English, and the deficiency of evidence for Powhatan's tribes in Virginia is particularly deplorable. Given Opechancanough's intelligence, stern foreboding, and sophisticated knowledge about Spaniards, as well as his familiarity with certain English ships in the Chesapeake in or after 1603, it was probably he who stationed sentinels—somewhat after the Spanish fashion—from Cape Henry to the York to warn of the approach of any white men in great ships. In the light of his great stature, prowess as a warrior, and rank as a king, it is almost certain that he was prominently involved in the elimination of the Chesapeakes and the white men from Roanoke—his *second* massacre. If this be true, then Opechancanough could actually have witnessed the entry within the Virginia capes of the three vessels of Captain Christopher Newport in April 1607.[25]

The first face-to-face meeting of the English invaders with Opechancanough occurred toward the end of May when Captain Newport made an exploring trip to the James as far as the falls. Gabriel Archer reported that on the return to Jamestown, the party stopped off at the village of "the King of Pamaunches," who was, if our reasoning is correct, the former Don Luis, now called Opechancanough. "This king (sitting in the manner of the rest)," the Englishman wrote in his journal, "so set his countenance, striving to be stately, as to our seeming *he became* [a] *fool*. We gave him many presents, and certified [to] him . . . that we were professed enemies of the Chessepians, and would assist King Pawatan [his brother] against the Monocans; with this he seemed to be much rejoiced." Little did these Englishmen suspect what manner of man they were dealing with; that this apparently savage clown was a wise and subtle leader with a wide knowledge of Europeans, had been at the Spanish court in the 1560s, and had a good understanding of the Spanish tongue. For the time being, in order to divine more of their motives, Opechancanough kept his counsel and assured the strangers of his peaceable intentions.[26]

Early in December 1607, Captain John Smith and a small trading party left the settlement at Jamestown in search of corn and other provisions. Ascending the Chickahominy River, they were surprised and captured by Opechancanough and two hundred warriors. Taken before the mighty werowance, Smith, fully expecting to be killed, played for time by presenting the chieftain with a round, ivory double compass dial and explaining its use. The rank and file of the natives were astonished at the action of the needle. Feigning surprise and curiosity, Opechancanough said nothing, apparently having elected for the time being not to reveal his familiarity with the transparency of glass and the magnetic needle. Instead, the king shrewdly drew out the loquacious and overweening prisoner, letting him "discourse of the sunne, moone, starres, and plannets." Later on Smith commented that "the King tooke great delight in understanding the manner of our ships, and sayling the seas, the earth and skies, and of our God. What he knew of the dominions [of Virginia] he spared not to acquaint me with, as of certaine men at a place called Oconahonan [in North Carolina] cloathed like me." 27

The compass episode had delayed Smith's fate less than an hour, after which the Indians tied him to a tree and as many as could stood there ready to shoot him. "But the king holding up the Compass in his hand" gave no sign, whereupon "they all laid downe their Bowes and Arrowes" and led him off to their village in triumph. Thus, for unvoiced reasons of state, Opechancanough spared John Smith's life, approximately three weeks before his niece, Pocahontas, again saved the Englishman in the presence of her father the Powhatan. 28

It is significant that when the prisoners were taken before Powhatan, he too was ready to release John Smith and his men, for the Indian brothers both knew that they could not expel the strangers. Even though they might be able to muster sixteen hundred fighting men as opposed to the small number of white men, firearms gave the latter a great military advantage. Consequently, Powhatan ordered that the Englishmen be set free. Henceforth the chieftains would bide their time and keep on the alert for the arrival and departure of Englishmen. 29

The inaccurate assessment of the Indian character made by Captain John Smith blinded him to the deep resentment of Powhatan, Opechancanough, and their fellow kings at the white men's presumption that they were entitled to settle on the Indians' land

and, in addition, order them to supply the English with corn. A prime instance of this arrogance occurred in January 1609 when John Smith, George Percy, Francis West, and thirteen others, having failed to persuade Powhatan and his people to trade their corn, sailed up the Pamunkey to deal with Opechancanough at that king's house. The presence of fifty or sixty warriors in the house alerted Smith to his dangerous situation, but undaunted, the Captain boldly seized the initiative and "in plain terms told the king this: I see Opechancanough your plot to murder me, but I fear it not." Because of the disparity in numbers, the Englishman challenged the huge Indian to meet him naked in a hand-to-hand combat on an island in the river: "The Conqueror take all." Caught off guard by such an unheard of proposal made in the presence of many of his leading tribesmen, the embarrassed Opechancanough tried to draw the Englishmen out-of-doors where two hundred of his men were awaiting the signal to kill them. Realizing the true intent of this maneuver and "vexed at that coward," Captain Smith (so he wrote later) did "take this murdering Opechancanough . . . by the long lock of his head; and with my pistol at his breast, [Plate 2] I led him [out of his house] amongst his greatest forces, and before we parted made him [agree to] fill our bark with twenty tuns of corn." Then the small but tough soldier turned to the assembled Pamunkeys and declared that he had had enough of their insolencies, and if they did not at once freight his ship and trade as friends, he meant to lade it with their "dead carkases." Years later Smith boasted about how the king trembled, nearly dead with fear, and that his followers were aghast that anyone would dare treat this mighty and enormous man thus. But Smith got away with it all.[30]

Apparently this fearless display of arrogance, backed by European firearms, produced the hoped-for result. Twice, however, Smith had humiliated Opechancanough in front of his people: first when he made him appear frightened and uncertain when he proposed single combat and again, by taking him by the hair lock, making him appear to be an abject coward. To these public insults he added a mighty threat to the entire Pamunkey tribe. Neither the king nor his loyal followers would ever forget such shame, and eventually they would make those uninvited strangers pay dearly for their actions.

C.Smith taketh the King of Pamavnkee prisoner 1608

Captain John Smith Threatening Opechancanough, 1608

Not long after these events, Pipsco, the principal werowance
living south of the James River, detracted still more from Opechancanough's public "image" by stealing away his "Chief woman."
Anyone familiar with the *Iliad* must have been reminded of Mene-
laus. William Strachey exclaimed that he had never seen "so handsome a savadge woman," who "with a kynd of pride can take upon
her a shew of greatnes." He thought that this tawny Helen in her
attire of feather and flowers appeared as "debonayre, quaynt, and
well pleased . . . as a daughter of the House of Austria, with all
her Jewells." Pipsco was promptly deposed by Powhatan from his
office, but nevertheless, whenever he went abroad—and he
usually showed up in Jamestown two or three times during a
summer—he was accompanied by "his best beloved," which, of
course, was most humiliating for Opechancanough.[31]

The kidnapping of Pocahontas, the favorite daughter of Powhatan, by Captain Samuel Argall in the spring of 1613 further exacer-
bated relations between the two races in Virginia. His action was,
in truth, the first of a long train of events that led irrevocably to
the precipitate decline of Powhatan's authority, to the rapid ascen-
dance of Opechancanough over the tribes of the Tidewater, and
straight on to the "Massacre" of 1622 with its tragic aftermath. Our
modern preoccupation with ideologies and social forces often
blinds us to the fact that it is men, not isms or movements, that are
most often the determining factors in history. What occurred in
1614 was not a new departure but rather a return to and intensify-
ing of the policy by which the Powhatan Empire had been formed
since 1571—the principles of which were preserved in the "prophecies." It was Opechancanough, a rising native leader both great
and wise, who fatefully dominated relations between the English
and the Indians from 1614 to 1622, and it was he whose estimate of
the situation turned out to be astonishingly accurate.

For almost a year the comely young woman who had saved the
life of Captain John Smith in 1608 was detained as a hostage at
Jamestown. In March 1614, an expedition headed by Governor Sir
Thomas Dale set out for Powhatan's seat at Werowocomoco on the
York to press for an exchange of the Princess Pocahontas for several runaway Englishmen, some stolen weapons and tools, five
hundred bushels of corn, and a guarantee from "the Great King" to
"be for ever friends with us." The only alternative for Sir Thomas

was "present warre." On the way up the river the English met with such "a great bravado" from the tribesmen that they felt warranted in going ashore and burning forty houses, making "freeboot and pillage," and killing five or six of the "Naturalls." Upon reaching Matchcot, not far upstream from Powhatan's chief residence, they observed about four hundred warriors assembled there; nevertheless, taking Pocahontas with them, they went ashore to parley. Two of her brothers desired to see their sister and, finding that she had been well treated at Jamestown, declared they could undoubtedly persuade their father to redeem her and "conclude a firme peace."[32]

John Rolfe and Master Robert Sparkes accompanied the sons to the aged Powhatan's house to negotiate with him, but he refused them admission to his presence, whereupon they were taken up the Pamunkey to talk with Opechancanough. The Indians pointedly said that "what he agreed upon and did the great King would confirme." As Powhatan's "successor," and already in "command of all the people," Opechancanough promised to do his best to effect the exchange and to use his influence for a truce. Even so the white men warned him emphatically that "if final agreement were not made" before harvest time four months thence, they would return and either take away (plunder) or destroy the Indians' corn, burn all the houses along the Pamunkey River, leave not a fish weir standing, nor a canoe in any creek thereabout; and that they would kill as many of the natives as they could. After this uncompromising announcement, the English sailed homeward to plant their crops.[33]

Here was *an unconditional ultimatum*, not the mutually agreed upon negotiated peace that Ralph Hamor, John Rolfe, and later historians reported. Further, Pocahontas provided no comfort for her aged parent. Soon after the event, Dale wrote: "The King's daughter went ashore, but would not talke to any of them, scarce to them of the best sort, and to them only, [saying] that if her father had loved her, he would not value her lesse than old swords, peeces, or axes; wherefore she would stil dwel with the English men, who loved her." It was not so much that she was forsaking her race and people as that she wanted to marry John Rolfe.[34]

Powhatan was near the end of the trail. His popular and vigorous brother had almost completely replaced him in authority, and his darling Pocahontas was determined to "turn English" and Christian. He also knew full well that the tough, white enemy would return in force at the end of August. It is clear that for reasons of his own and of high tribal policy, the calculating Opechancanough strongly urged on him the wisdom of giving in to the English demands. Powhatan permitted Pocahontas to return to marry Rolfe and live at Jamestown (a month later, however, he refused to bestow another daughter on an importunate Governor Dale), and he did seek peace. He asked Hamor in May 1614 to tell Sir Thomas that "though he hath *no pledge at all*, he need not distrust an injurie from me or my people; there have beene too many of his men and mine slaine, and by my occasion there shall never be more . . . I am now old and would gladly end my days in peace." "More for feare then for love," Powhatan personally accepted a forced, unwritten peace, one that his people regarded more as a truce.[35]

When the English were threatening the Indians at Pamunkey, Ralph Hamor had informed Dale of the desire of John Rolfe to marry Pocahontas. The governor's approval of the interracial union, wrote Hamor, "was the onely cause hee was so milde" with the natives, otherwise he "would not have departed their river without other conditions." When Powhatan heard of the proposed marriage, he had virtually no choice but to give "his sudden consent." Ten days later he sent Opechancanough "to give her in the Church" at the ceremony that took place about the first of April 1614. It seems very likely that it was Opechancanough who proposed that he, accompanied by two sons of Powhatan, go to Jamestown: it would give the former Roman Catholic a chance to see how Protestants worshiped, an opportunity to spy out the defenses of the little village, and his presence at the ceremony would also help his humiliated brother to save face. Furthermore it would enable him to strengthen his newly formed friendship with Sir Thomas Dale and pursue the new course begun at Pamunkey of ostentatiously professing friendship for the English.[36]

The marriage of Pocahontas and John Rolfe was not the fundamental reason for the truce; it was, however, an acceptable gesture toward it. Sir Thomas Dale left no doubt of this when he wrote to London on June 18: "Now you may judge . . . if the God of battailes have not a helping hand in this, that having our swords drawn, killing their men, burning their houses, and taking their corne: yet they have tendred us peace, and strive with all alacrity to keep us in good oppinion of them." About the same time the Reverend Alexander Whitaker praised "our religious and valiant Governour": "he hath brought them to seek for peace of us, which is made and they dare not breake." To Dale and the rest of the colonists, the benefits secured were the return of their arms and tools, the safety of their cattle, an ample supply of corn, and greater familiarity with the Indians. They saw the marriage only as "an other knot to bind this peace the stronger."[37]

There was no written treaty with a specified list of articles guaranteeing the truce forced upon Powhatan and his chiefs, and each side thought of it differently. In *A True Discourse,* composed not long after the confrontation on the Pamunkey, Ralph Hamor gave the English view of "our established friendship with the Naturalls . . . which I hope will continue so long between us, *till they shall have the understanding to acknowledge how much they are bound to God for sending us amongst them* (than which) what work would be more acceptable to God, and more honorable to our King and country." In this unconsciously fatuous formulation of the theory of "Assmiliation," Hamor reveals how completely the red men and white misunderstood each other. The Virginia colonists grossly overestimated what they had gained by humiliating Powhatan in this personal capitulation, and his brother Opechancanough never forgot it.[38]

"Hearing of our concluded peace with Powhatan, as the noise thereof was soon bruited abroad," Hamor believed, was what impelled the Chickahominies to seek peace with the white men, who had been their enemies since 1607. But the real story was somewhat different. Next to the Pamunkeys in numbers, they were a lusty and daring tribe with at least five hundred fighting men and had long lived independent of imperial control by Powhatan. Their strategic location between Jamestown and the Pamunkey country suggested to Opechancanough the creation of a buffer zone to keep the white men away from the lands of Powhatan. Capitalizing upon their well-known fear of Powhatan, who was now apparently in league with the English, Opechancanough astutely persuaded the Chickahominies to abandon their hostility toward the settlers from overseas. Governor Dale, after ignoring three or four applications from the anxious natives, sent Captain Argall to negotiate with them. In 1614 he concluded a formal treaty containing five specific provisions by which the members of the tribe recognized "as their supreame head" the governor of Virginia as the deputy of King James of England. They even changed their name to Tossantessas (Englishmen) and accepted red-coat "liveries." In this instrument of friendship and alliance, Article 3 provided that these Indians would aid the English against any attack by the Spaniards, "whose name is odious among them," or any other tribes that might attack them. Article 5 obligated the Chickahominies to pay into the store-house at Jamestown annually a tribute of two bushels of corn for each of their five hundred warriors.[39]

By adopting these stratagems, Opechancanough, whose goal was the traditional one of uniting the tribes of Virginia against the strangers, made use of his brother's failure to initiate a positive program for his people. First, he succeeded in deluding the English into thinking that he was their friend and that peace between the races would be permanent. Second, and more significant, he induced the recalcitrant Chickahominies to ally themselves with Jamestown, thereby increasing the false sense of security that prevailed there. By these achievements he showed himself to be a supreme master of forest diplomacy.[40]

The intrigues of Opechancanough with the Chickahominies revealed a serious rift in the tribal command of the Powhatans. Ever the realist, he had persuaded them to make peace with the English, but in doing this he unintentionally (or so it seems) aroused apprehension in his failing brother, who had always looked upon that large tribe as unruly, one that he could never subject to his authority. On the other hand, the Chickahominies, grateful for Opechancanough's concern and counsel and having no principal chief, made him their ruler, "the King of Ozinies." Convinced that Opechancanough had "debauched" the Chickahominies, Powhatan grew increasingly suspicious that his brother had "entered into a Conspiracy to betray him" into the hands of the white men. Sometime between June 1614 and March 1616, "that subtill old revengefull Powhatan" determined to alter the succession by designating his lame second brother Itopatin as the next Powhatan. If

Opechancanough knew of this at the time, he said nothing but awaited the outcome. Meanwhile he worked away patiently and confidently on his secret plans.[41]

In April 1616, accompanied by the Rolfes and about ten Powhatan Indians, Sir Thomas Dale returned to England. He left as his deputy George Yeardley, a former soldier in the Netherlands who had come to Virginia in 1609. The new governor promoted the growing of tobacco to the exclusion of other crops with the result that the colony's food supply was soon exhausted. Early in the summer he decided to send to the Chickahominies for the "tribute corn" required annually by the treaty of 1614. Prompted by their new king, they returned "such a bad answer" that the English were insulted. Unaware of Opechancanough's role in the dispute, Yeardley turned to him for advice and was told that the Chickahominies had killed a number of much needed cattle and swine.[42]

Still playing a double game, the chieftain urged the governor to march into the buffer region and force a parley, while, at the same time, he shrewdly pressed his "subjects" to resist Yeardley's demand. Two or three hundred natives assembled on the banks of their river to meet the governor who arrived with about a hundred armed men. Derisively calling him "onely Sir Thomas Dales man" and treating him "with much scorne and contempt," the Indians insisted that they had paid the tribute and would not obey him "as they had done his Master." Instead of reasoning with them, the irate official ordered a volley of shot fired which killed between thirty and forty tribesmen. The English took twelve prisoners, whom they released after settling for one hundred bushels of corn as a ransom—which was never paid.[43]

Stopping at Ozinies on his homeward journey down the Chickahominy, George Yeardley encountered Opechancanough who, "with much adoe," explained how he with great pains had procured the peace between the English and the Chickahominies and, "which to requite, the natives called him King of Ozinies, and brought him from all parts many presents. . . ." The Englishman obviously failed to grasp the import of what he had just heard for the first time, for once back at Jamestown, he and his men, the other inhabitants, and all other "Savages" lived together "as if wee had been one people all the time Captain Yearly [sic] staied with us." This was made possible "because of the great league we had with Opechancanough."[44]

The "Great King" had lived up to the description "wily" and "crafty" given to him by contemporary commentators, for he had completely outwitted the English. As Captain John Bargrave recalled the massacre of the Chickahominies, he agreed that "this perfidious act made them all flye out and seek Revenge: they joined with Opichankano and having by stealthes and murthers diverse tymes afflicted such of our Colony as they could meet with and dali offered them wronges. No Revenge was taken, but all was putt up with in so much that before the last Massacre [1622] our Colonyes were almost made subject to the Savages. . . ."[45]

With our modern gift of hindsight it is evident that, from the time the English landed at Jamestown in 1607, sooner or later a clash between them and the natives was inevitable. The wanton slaughter of his Indian allies ordered by Governor Yeardley in 1616 foreshadowed the famous "Massacre" of the English in 1622. Opechancanough never forgave the governor for his senseless action and frequently in the years ahead manifested his hatred and contempt for him. Nevertheless he profited from the incident, for the unfortunate Chickahominies moved voluntarily and permanently into the Powhatan camp. As a result Opechancanough was free to solicit support from, if not alliances with, the tribes on the Rappahannock, the Potomac, and over on the Eastern Shore of Virginia, and during the next six years he and his people grew progressively stronger and bolder.[46]

Meanwhile Powhatan was steadily losing influence among his tribes. Lacking Opechancanough's first-hand knowledge of Europeans, the naïve old chief resorted to an unbelievable scheme for learning more about the strength of the English in their own land. He sent Tomocomo, his ablest and most reliable supporter, in the ship with Dale in 1616 to determine how much corn the English had and the size of the population; he was to record the latter by cutting a notch in a stick for each person he saw! His experiences on the first day after landing at Plymouth overwhelmed him, and he threw the stick away. And in the remainder of his stay he developed an abiding dislike of all Britons and their land. In May 1617 he returned to Virginia with the new governor, Samuel Argall, who immediately sent him to invite Opechancanough to visit

Jamestown and receive the presents intended for him. Just as grat-
ifying to the chief was to be present with "his great Men" when
Tomocomo railed against England and his former friend Sir
Thomas Dale, only to have his charges completely disproved. Thus
the most dependable of Powhatan's chiefs was publicly "dis-
graced," and the chieftain suffered the loss of most of the small
amount of prestige and authority remaining to him. Shortly after,
the aged chieftain, after delegating his power to his brothers Opechancanough and Itopatin jointly, went away to live with the King
of Mayumps on the Potomac. In April 1618 he died; by his final
act, Itopatin, his lame second brother, succeeded him as the new
Powhatan.[47]

Times were bad for the Indians, as Argall discovered when he
arrived to replace Yeardley. The summer of 1618 was one of very
high mortality among the tribes, and that contributed to their poverty and inability to pay tribute. Furthermore murrain among the
deer had reduced the meat and leather supply of the natives and
consequently their trade. Itopatin and Opechancanough, seeing
the English both flourishing and greatly increasing in numbers, ex-
pediently renewed the personal peace made by Powhatan in 1614.
Nevertheless Opechancanough, unwilling to play second fiddle,
deftly exploited his personal popularity with all of the Indians
(which Itopatin lacked) in a struggle for power, and within twelve
to eighteen months he had "grasp'd all the Empire to himself" and
become *the Powhatan.* Absorbed in Indian matters of state and
hating Yeardley, who had again become governor in April 1619,
Opechancanough declined to meet with the officials at Jamestown.
Such a refusal, John Rolfe reported with some concern, "causes us
[to] suspect his former promises."[48]

During Opechancanough's contest for leadership of the Tidewater tribes, the relations of the natives with the English, already
strained, steadily worsened. A shortage of powder had caused Governor Argall in May 1618 to prohibit any shooting save in self-
defense, and to forbid all private trade with the Indians, the em-
ployment of native hunters, or instructing them in the use of guns
"lest they discover our weakness" before the arrival of a new supply. Noticing that there was no firing of guns, the Powhatans and
Chickahominies concluded that "our peeces were, as they said,
sicke and could not be used." Early that summer the Chicka-

hominies boldly killed Richard Killingbeck and four others who had gone up their river to trade secretly with them, saying that it was in revenge for their brethren whom Yeardley's men had slain. A week later they murdered one Fairfax, his wife, two children, and a youth. Governor Argall demanded satisfaction from Opechancanough, but he, claiming ignorance of the whole affair, excused it. When some of the tribesmen expressed fear of retaliation by the English, the chieftain, to calm them down, artfully told the governor that he hoped the men of Jamestown would not revenge the killing by "those fugitives" on the innocent people of Chickahominy town. In addition he sent a "basket of earth" to symbolize Argall's possession of the place and promised that as soon as he apprehended the culprits, he would send their heads to him for his satisfaction; "but he never performed it." He also assured the English that "the peace should never be broken by him." Governor Argall made no attempt to punish the Chickahominies for what the Company in London branded an "outrage," but it resulted not from inaction or restraint on his part but from the lack of powder.[49]

The year 1619 was a year of decision. At London a new administration of the Virginia Company, led by Sir Edwin Sandys, sought to reinvigorate the colony by granting the people self-government and sending out settlers in unprecedented numbers. It also encouraged the establishment of large estates, or "particular plantations," and provided for a grant of fifty acres of land (a headright) to every person transported there after 1618 who remained for three years. Such a policy necessitated both a dispersal of settlement and the acquisition of large tracts up the valleys of the James and Chickahominy.[50]

Actions taken to implement these policies permitted no turning back by either race. The officials in London dispatched ominous instructions to Sir George Yeardley on June 21, 1619: "The outrage don by the Chekohomini deserveth a sharpe revenge . . . not only to the personall destruction of the murtherers, but *the removing of that people further of [f] from our Territories by all lawfull meanes* if the same be not allready don by Captain Argall, as he seemeth to insinuate. But for the rest mainteyne amity with the natives, soe much as may be, and procure their Children in good multitude to be brought upp and to worke amongst us."[51]

Whether or not Opechancanough ever learned of this early removal proposal, he knew all about the other part of English Indian
policy: assimilation of the Indian by conversion to Christianity,
religious education of his children, more or less forced labor, and
other "civilized" devices. They were much like the methods he
had seen the Spanish using in Mexico many years before. As far
back as 1609 Sir Thomas Gates had been advised by the Company
that conversion would be best achieved by procuring from the natives "some convenient nomber of their Children to be brought up
in your language and manners." [52]

Not much had been done to convert adult Indians or instruct
their children before 1617. On his return from England that year,
John Rolfe reported to Sir Edwin Sandys that "the Indyans are
very loving and willing to parte with their children." Such enthusiasm was premature, however, for in March 1619 Governor Yeardley was notified by London that to promote a proposed treaty with
Opechancanough "touchinge the better keeping of the Infidells
Children which are to be brought upp in Christianytie," he should
select something from the Magazine at Jamestown as a gift for "the
better attayninge their ends of him." In reply Sir George frankly
warned that "the spiritual vine you spake of will not so sodaynly be
planted . . . the Indians being very loath upon any tearmes to
part with theire children." In his opinion the best contrivance
would be for Opechancanough to appoint from his tribes a number
of families—one for every corporation and particular plantation—
and for the English to provide each of them with a house and a
plot of ground for planting corn. The Powhatan has consented to
this, Yeardley wrote, and by such means both parents and children
can live together, receive instruction, and raise their own food at
no cost to the settlers. [53]

Incidents occurred in 1619 that illustrate vividly the mounting tension between the two races. The first took place in the
spring when Ensign Harrison was returning in one of Captain
John Martin's shallops from an unsuccessful voyage up Chesapeake
Bay. Seeing some natives coming out of a creek into the York
River in a long canoe laden with corn, the English insisted that
the Indians sell them their cargo. When they refused to comply,
the English took it by force and paid them in "trucking stuffe."
The Powhatans made their complaint known to Opechancan-

ough, and he demanded that Governor Yeardley "give them justice." The case came before the first legislative assembly on July 30, and the members, aware that "such outrages might breede danger and loss of life to others of the Colony," censured Captain Martin mildly, ordering him to give satisfaction for his offense and, in the future, to secure permission from the governor before trading with the Indians.[54]

The members of the Assembly were called upon in August to deal with what they assumed was strictly a white man's affair. Robert Poole was an interpreter who carried messages for Governor Yeardley to and from Opechancanough. He persuaded Sir George that if he would promise to receive Opechancanough, "the greate King" would come to visit him in Jamestown. This occurred at a time when Yeardley was extremely uneasy about the peace with the Powhatans and Chickahominies, for the colonists had begun to spread out to "many struggling Plantations" recently weakened by "the great mortality." He therefore sent two Englishmen to Opechancanough to assure him of compliance with his wishes, but they were promptly dismissed and sent back with the "frivolous" answer that "he never had any interest to come."[55]

This same "very dishonest" Robert Poole, according to John Rolfe, also asked him and Captain William Powell to go to Pamunkey ("for Opechancanough professeth much love to me and giveth much credite to my wordes"). The governor approved, and the two of them went up the river in a shallop to within five miles of the chief's village without being observed. Their early arrival caught the Powhatans off guard, as did that of an English frigate soon after, "which much daunted them and put them in great feare" of a clash. Perhaps it was because of this that the two messengers delegated by Rolfe to meet with Opechancanough were at first treated harshly. The chieftain wanted to know why Rolfe himself would not leave the shallop. After the couriers explained that Rolfe had the ague and delivered Yeardley's conciliatory message, Opechancanough was gracious, and the party was sent homeward "in great love and amyty." As it turned out so satisfactorily, the white settlers were much relieved by the way both sides handled the crisis created by Poole.[56]

About this time also Robert Poole charged under oath that at Opechancanough's Pamunkey "courte" another interpreter, Henry

Spelman, "spake very irreverently and maliciously against" Governor Yeardley. Called before the Assembly, Spelman denied most of the charges, but he did admit that he had told Opechancanough that a new governor, one greater than Yeardley (possibly the Earl of Warwick) would come within a year. In the view of the members of the Assembly, Spelman had "alienated the minde" of the natives from the governor by this and other reports and brought him into disesteem "with Opechancanough and [placed] the whole colony in danger of their slippery designes."[57]

We shall probably never find out how much Opechancanough knew about, or was responsible for, Poole's machinations. John Rolfe reported that Poole "even turned heathen" but in spite of that maneuver, he lost favor with the tribesmen and was accused and condemned by them of being "an instrument that sought by all the meanes to breake our league." As for the white men, the governor and his advisers continued to question the durability of the "league" and revealed their doubts in another incident of 1619.[58]

On November 11, 1619, Sir George Yeardley sought the advice of his Council about a "project" submitted to him at Charles Hundred on October 25 by Nemattanow, the representative of the Powhatan, who was "an Indian Comonly Called . . . English Jack with the fethers." Opechancanough wanted eight or ten armed Englishmen to assist him in a battle of revenge against some warriors living beyond the Falls of the James who had murdered certain Powhatan women "contrary to the law of Nations." He offered the English an equal share of all "the booty of male and female children," corn, and lands. The councilors were willing to agree to the proposition because "they found the warre to be lawfull and well grounded," required little aid on their part, and significantly was "not of Consequence enough for Opachancano to put any treacherous disaster upon." This, they said, was "only to oblige him, who ever since Sir George Yeardlies coming in hath stood aloof upon termes of dout and Jealousy and would not be drawne to any Treaty," notwithstanding "all the Arte the governor could use." The members also thought that such aid might win "amity and Confidence from Itoyatin," who was still "the great Kinge" to them, as well as his powerful brother, and "their subjects."[59]

The Council particularly welcomed the prospect of obtaining Indian children who would live among them and, it was hoped, promote assimilation. Such a program had been recommended to Governor Gates in 1609 and again in the instructions sent to Yeardley, but both the Powhatans and Chickahominies had been adamant in their refusal to send any of their children to the new college at Henrico.[60]

When Sir Edwin Sandys and his supporters took over the Virginia Company in 1618, they proceeded rapidly and with evident sincerity to resurrect and carry out the policy stipulated in the charter of 1606. By it, the colonists were enjoined to propagate "the Christian religion to such People as yet live in Darkness and miserable Ignorance of the true Knowledge and Worship of God," and to bring "human Civility and a . . . settled and quiet Government" to them. The Company also emphasized in its instructions to the first settlers that "you must have great care not to offend the naturals, if you can eschew it."[61]

We have observed how the settlers and their leaders callously or unwittingly ignored the orders from the Company to treat the natives humanely and to try to convert them. Generally the English looked down upon the "tawnyes" as being little more than brutes, and the Indians did not fail to recognize the posture as abasing.

A few colonists, conceding this failure to establish good relations with the natives, gave the Indians high marks for their talents and perspicacity. Reporting his own personal observations in *Good Newes from Virginia* (1613), the Reverend Alexander Whitaker severely chided his compatriots for their attitudes: "Let us not think that these men are so simple as some have supposed them: for they are of bodie lusty, strong, and nimble; they are [an] average understanding generation, quicke of apprehension, sudden in theire dispatches, subtle in their dealings, exquisite in their inventions, and industrious in their labour. I suppose the world hath no better marks-men with their bow and arrowes then they be." And with rare discernment, Master Whitaker also pointed out that "there is a civill government amongst them, which they sturdily observe, . . . they [also] observe the limits of their owne possessions, and incroach not upon their neighbour's dwellings."[62]

The year 1620 passed without any serious disturbance, and its successor seemed to dawn auspiciously, at least for the white men. The officials of the Company learned from the Council in Virginia during the spring that "whereas before they had ever a suspicion of Opechancanough, and all the rest of the Salvages, they had an eye over him more then any; but now they all write so confidently of their assured peace with the Salvages, there is now no more feare nor danger either of their power or trechery; so that every man planteth himselfe where he pleaseth, and followeth his businesse securely."[63]

There may have been no actual clashes between the natives and the English, but one Englishman immediately sensed that there was no love lost between the two. This man, Master George Thorpe, a stockholder of the Company, crossed to Virginia in January 1621 and, as befitted his rank, was immediately admitted to the Council. The Indians never had a more understanding or better friend than this compassionate man from Gloucestershire. In May he wrote from the new college at Henrico to Sir Edwin Sandys that he believed God was displeased with the English for neglecting to convert the heathen: "There is scarce any man amongst us," he asserted, "that doth soe much as affoorde them a good thought in his hart, and most men with theire mouthes give them nothing by maledictions." They are falsely persuaded that "these poore people have done unto us all the wrong and injurie that the malice of the Devill or man cann afford, whereas in my poore understandinge, *if there bee wrong on any Side it is on ours,* who are not so charitable to them as Christians ought to bee."[64]

Not long after his coming to Virginia, Thorpe began visiting Opechancanough at the latter's request to confer about religion. The American's knowledge and grasp of Christianity proved amazing: on one occasion the pagan confessed, "moved by naturall Principles, that our God was a good god, and better much than theirs, in that he had with so many good things above them endowed us." The Englishman accepted as genuine the Indian's request to be "instructed" in the Anglican faith and looked forward to converting him. Even more startling for historians today was the discovery that Opechancanough had "alsoe . . . some knowledge of many of the fixed starrs, and had observed the north Starr and the course of the Constellatione about it, and called the greate beare Manguahaian, which in theire Language signifies the same." Here again one detects in Opechancanough's conversation a somewhat sophisticated grasp of things Christian and European that could scarcely have been acquired in Tidewater Virginia from fellow Algon-kians.[65] In keeping with his own views, George Thorpe labored hard at converting these pagans: when any of his underlings "did them the least displeasure," they were severely punished. He even had some English mastiffs killed because the natives were afraid of them. Finding Opechancanough, the new Powhatan, living in a hovel, or "hog-sty," made with a few poles and stakes covered with mats "after their wild manner," Thorpe built a "fayre" English house for him in the hope that it might civilize him. What particu-larly delighted Opechancanough was the lock and key, "which he so admired, as locking and unlocking his doore an hundred times a day; he thought no device in all the world comparable to it."[66]

In marked contrast to George Thorpe was the Reverend Jonas Stockham, a newly arrived Anglican minister, who complained about the Company's policy toward the Indians on May 28, 1621: "I can finde no probability by this course to draw them to good-nesse; and I am perswaded if Mars and Minerva goe hand in hand, they will effect more in an houre, then those verbal Mercurians in their lives; and till their Priests and Ancients have their throats cut, there is no hope to bring them to conversion." Most Virgin-ians agreed with Master Stockham and, as was the case ever since 1607, the harvest of converts was "slender." As a matter of fact, religion was always a peripheral rather than a central issue be-tween the Indians and the English because so little was actually done about it.[67]

Instructions sent in July 1621 to Sir George Yeardley, the out-going governor, re-emphasized the need "to draw the better dis-posed of the Natives to Converse with our people and labor amongst them with Convenient reward, that therby they may growe to a liking and love of Civility, and finally be brought to the knowledge and love of god and true religion, which may prove also of great strength to o[u]r people against the Savages or other In-vadors whatsoever; and they may be fitt Instruments to assist af-terwards in the more generall Conversion of the Heathen which wee so much desire."[68]

At the same time that these admonitions were given, the gov-ernor and all officials in the colony were advised to take care "that no injurie or oppression bee wrought by the English against any of

the Natives . . . wherby the peace may be disturbed and ancient
quarrells (now buried) bee revived, *Provided nevertheless* that the
honour of our Nation and safety of our people be still preserved
and all manner of Insolence committed by the natives be severely
and sharpelie punished." The inconsistency and irony of these instructions seem not to have occurred to contemporaries.[69]

So convinced were the white men that they were conferring
immeasurable benefits upon the red men that they never recognized or sensed the legitimate resentment of Opechancanough and
his people at the takeover of their lands and produce. Such evidence as survives shows that the Indians opposed the whites from
the very beginning with determination and skill and seldom relaxed their pressure. "The Indians," wrote Robert Beverley, who
knew them well, "never forget nor forgive any Injury, till satisfaction be given, be it National or Personal; but it becomes the business of their whole Lives, and even after that, the Revenge is en-
tail'd upon their Posterity, till full reparation be made." And, as
we have noted, there were grievances and insults, as well as at-
tacks and killings, to justify their animosity.[70]

Despite underlying tension on both sides, the Indians profited
from a growing trade in corn and other commodities with the English during the "four years of security" after 1618. The Council
considered, and so reported to the Company in London, that besides prospects for a great trade in corn there should be a "good
meanes also for converting" the natives to Christianity and drawing
them "to live among our people." As Captain John Martin testified
later, this traffic made it possible for Opechancanough to employ
many "auxiliaries" in a way that Powhatan never could; here again
he exhibited an un-Indianlike prescience by acting much more like
the white Europeans than his red brothers.[71]

The remarkable success of his tribal diplomacy and the mounting strength of his fighting forces inspired Opechancanough with
great confidence in his capacity to deal with the English. To allay
their suspicions while continuing his plans, he unexpectedly made
"a peace and league" with Jamestown early in 1621. He must have
been genuinely gratified by the spectacle of red men and white
mingling together daily in trading and social pursuits at the settlements.[72]

On the other hand, the influx of settlers from England understandably alarmed him. From 1607 to 1617, the white men intruded only on a circumscribed part of the Indians' homeland. After 1619 the population rose rapidly from one thousand to twenty-three hundred according to the Company, and there was no reason to think their numbers would decrease. This more than anything else augured ever more encroachments on tribal lands. The new course of allocating tracts to favored Englishmen and the unrestrained scattering of plantations over the Peninsula menaced the whole basis of Indian society. Proof of this threat was revealed in the comments of the Company about a grant made by Governor Yeardley to Sir Edward Barkham between 1619 and 1621. The Londoners considered it very "prejudicial" and "dishonourable" to their Company "in regard it was lymitted with a Proviso to compound with Opechankano, whereby *a Soveraignity in that heathen Infidell* was acknowledged, and the Companies Title thereby much infringed." One would like to know more about this matter.[73]

For some years it had been the custom of Opechancanough to make an "annual progress through his pettye provinces," William Powell informed Sir Edwin Sandys in April 1621. That year when he reached the Potomacs, because he was already or soon would become *the Powhatan,* he was entertained with "the greatest honour that Nation" could offer. A special feature of the ceremonies was a display by the young Potomac braves in "a souldyerlike manner" of "their undaunted valours" before "his Majesty."[74]

Again, in September or October, when great numbers of Indians from all over the Chesapeake Basin were to be assembled at the ceremony for the "taking up of Powhatans bones," the great werowance plotted that afterward they should "sett uppon every Plantatione of the Colonie." At least this was a story told by the King of the Eastern Shore, no friend of Opechancanough, to Sir George Yeardley, and later, after the disaster of 1622, it was related to the Company in London. The governor promptly went in person to every plantation, took a muster, and charged the inhabitants to watch and ward alertly. "But Apochancano ernestly denying the plott, and noe aparant proofe brought in, our people fell againe to their Ordinary watch."[75]

Sir Francis Wyatt replaced Sir George in November 1621; he found the country quiet "and all men in a sense of security." Not only had the peace been solemnly ratified, but at the request of

Opechancanough it was "stamped in Brasse and fixed to one of his Oakes of note" at Pamunkey. Shortly after arriving in Jamestown, the new governor dispatched Captain George Thorpe with greetings to Itopatin and Opechancanough: the latter declared himself very much satisfied with the new governor's coming and readiness "to confirm their leagues" of peace. The brothers confessed that they had been "in some Jealousie whether our new Governor would continue the League or nott." Unexpectedly, in the light of past refusals, Opechancanough intimated he was willing that not only could some of his people go to live among the English but held out the possibility of several white families dwelling among his tribesmen.[76]

Upon assuming the office of the Powhatan, Opechancanough took a new name, *Massatamohtnock* (which did not signify a white soul). He was a native American patriot who had ample reasons for his deeds and deep resentment of the white men. Every act of the English since 1607 had been that of uninvited and unwanted outsiders relentlessly bent upon occupying the natives' domains and taking whatever they pleased in way of provisions. They forced the Powhatans and Chickahominies to trade and pay tribute; they demanded that Indian children be separated from their families and sent to them to be "civilized" and Christianized—their insistent claims for the superiority of the Anglican faith irked all but a handful of tribesmen. From 1607 to 1621 the English killed many more Indians than the latter killed white men. Moreover the large immigration of 1619–21 threatened their hunting lands and forced the abandonment of some Indian villages and presaged an eventual outnumbering of the aboriginal population at a time when assimilation—the Englishing of the heathen—could no longer be avoided.

All efforts at inducing the Europeans to leave Virginia having failed, it seemed that the prophecy—that a third body of strangers would defeat the Powhatan's people—was about to come true. Now in full control of thirty-two tribes and assured of the assistance of many "auxiliaries" from among his Algonkian allies along both shores of Chesapeake Bay, Opechancanough was confident that his people were, for the first time, ready to force a showdown with the English invaders. Since 1614 he had been making plans, winning formidable support for his cause, and succeeding in lulling the white enemy into security by a brilliant use of dissimulation. Opechancanough concluded that the juncture he had long planned for was at hand. The premature disclosure of his "plott" for an assault on the whites at the time of the Powhatan memorial ceremony caused him to wait for a favorable moment for such a strike. When, in the spring months of 1622 it arrived, he was ready.

The opening was provided unwittingly by Nemattanow, one of the fiercest and bravest of the Powhatan captains, but no planner, no counselor, certainly no diplomat, and well known to be a man "farr out of the favor of Apachancono." Just before the discovery of the Indian plot scheduled for the previous fall, this warrior, Jack of the Feathers, as he was known to the English, and some of his tribe killed several white settlers. In keeping with the treaty concluded earlier that year, Governor Wyatt and his Council offered to do justice to him "if uppon the taking upp of the dead bodies it might appear that Nenemachanew had no hande in theire deaths, which was all Apachancan[o] required." As part of his policy of delay, the Powhatan sent out for, or "fained to searche," for, the bodies.[77]

It is not difficult to understand why a chief of the caliber of Opechancanough placed little trust in Jack of the Feathers, for, other than being a "great Warriour," he showed little if any sense or judgment. He had convinced himself and most of the tribesmen that he was invulnerable to gunshot, in fact, immortal. Early in March 1621/22 he appeared at the plantation of one Morgan and induced him to go with him to the Pamunkey "to trucke." On the way he killed the white man and after several days returned to Morgan's place wearing his victim's cap. Jack so tried "the patience" of two youthful servants of Morgan, who inquired after their master, that they shot and fatally wounded him. While they were taking him in a boat to the governor, who was near by, the dying warrior pleaded with the boys not to tell anyone that he had been slain by a bullet and to bury him among the English so that his people would never discover that he was mortal.[78]

When the news of his death reached Opechancanough, he quickly sent word to Governor Wyatt that the death of Nemattanow "beinge but one man, should be no occasion of the breach of the peace, and *the Skye should sooner falle then [the] Peace be broken, one his parte*, and that he had given order to all his People to give us noe offense and desired the like from us."[79]

Here was just the kind of incident that the native leader needed. Jack of the Feathers, as he said, meant little to him, but to the general run of warriors he meant much: he was their great hero, and the English had murdered him. Cunningly, Opechancanough used Nemattanow's death to arouse all of his race in the Tidewater for the supreme effort, long in preparation, to rid their country of the white men once and for all. Perhaps better than any other living being in Virginia this man understood that two races with two radically different ways of life could not live together; it would have to be one or the other, and it must be now or never. It is clear that the shooting of Nemattanow was but the pretext for, not the cause of, the Indian uprising; it dated back directly to 1613 and, remotely, to 1607 or earlier.[80]

Notwithstanding the orders issued to the colonists to be constantly on the alert, the settlers had again lapsed into easygoing ways with the natives who, as the Council confessed in January 1623, by coming "daily amongst us and putting themselves in our powers, bread in our People a securitie." The Indians borrowed boats from the English to cross and recross the James River "to consult of the devellish murders that ensued, and of our extirpation." Shortly before the attack, one Brown, who had been living in a native village to learn the language, was sent back "in a friendly way" to Ralph Hamor. "And such was the conceit of firme peace and amitie," a contemporary wrote, "as that there was seldome or never a sworde worne, and a Peece seldome, except for a Deere or Fowle."[81]

Unexpectedly, a few days later, the sky did fall. Simultaneously, on March 22, 1622, the native Americans fell upon and butchered 347 white men, women, and children. Even George Thorpe, the great friend of the Indians, was murdered. Edward Waterhouse reported that the whole population would have been wiped out "if God had not put it into the heart" of Chanco, a Pamunkey servant in Surry, to disclose the plot to Richard Pace "who had used him like a son." Pace rowed across the James and alerted the governor, thereby saving the lives of the inhabitants of

Jamestown and several plantations roundabout; the Indians did set fire to several nearby houses and killed or drove away numbers of livestock.[82]

With attention to and concentration on the smallest details, Opechancanough had organized his own tribes and allied warriors for a *Blitzkrieg* to exterminate the white invaders. His achievement won a grudging admiration from many of the English leaders. George Wyatt of Kent, father of the governor and a close student of European warfare, acutely pointed out how the Indians' "Intelligences served them wel[l], and seeme to have dived far into some of our owen fals brestes. They assembled readily in great silence, proportioned and distributed their numbers to the partes of their charge with seconds to each." Combining both native and European techniques for surprise and deception, the attack was probably the most brilliantly conceived, planned, and executed uprising against white aggression in the annals of the American Indian. And it came within an ace of being a total success.[83]

This deliberate, marvelously timed attempt to root out all strangers in the land was, no doubt, thought of by the Powhatan and his race as a stroke for freedom, one that justified deceit, dissimulation, and the brutal, bloody methods employed. In the classic European sense, it was a *defensive war*, one provoked by the intruders from the British Isles, but the English termed it a "Massacre," and it is still so-called in the schoolbooks. In *A Declaration of the State of the Colony*, the official report of this "Theame of Sadnesse" published in London late in 1622, Edward Waterhouse referred to the Indians as "that perfidious and inhumane people" who slaughtered us "contrary to all Lawes of God and men, of Nature and Nations."[84]

All Englishmen who wrote about the attack agreed that the blandishments and seemingly generous actions, which Opechancanough used so successfully to lure the settlers into a fatal sense of security, were essential for the success of the plot. The colonists, said Robert Beverley, "became everywhere familiar with the Indians, eating, drinking, and sleeping among them." In this way the natives learned the strength of the English, the use of firearms, where to find their chosen enemies at all times whether they were at home or in the woods and whether they were in a condition of defense. The former Don Luis had learned a great deal more about the white man during his years in Mexico and Spain that just the religion of Rome.[85]

For several months Opechancanough apparently overestimated the completeness of his strike. At the end of March he had sent two baskets of beads to the King of the Potomacs and urged him to kill Captain Raleigh Crashaw, a trader, "assuring him of the Massacre he had made, and that before the end of two Moons there should not be an Englishman in all their Countries." He was still holding twenty English prisoners at Pamunkey in June when Captain Madison arrived with a message from Governor Wyatt requesting their release. To this the great chief gave an "insolent answer." About this time, too, in some fashion, he dishonored the picture of James I.[86]

In seeking to comprehend what had happened, the colony's leaders concluded, not unreasonably for their times, that the devil had instigated "this surprise." The more plausible explanation offered was that it was the natives' fear that the English were taking over their country and trying to alter their way of life that produced "the bloody act." Self-deception rather than hypocrisy was implicit in the English conviction that "our hands, which before were tied with gentleness and fair visage, are now set at liberty by the treacherous violence of Savages not untying the knot but cutting it." Herein we discover the first ironic principle of American history: that in the name of God, by right of war and the law of nations, the English were free to invade the natives' country and "destroy them who sought to destroy us."[87]

The question promptly arose whether the natives should be completely wiped out or merely reduced to a sort of peonage and employed for the white men's purposes as the Spanish had done. The majority of Englishmen, both in the colony and at London, cried out for revenge, ruthless and total. In *Virginia's Verger* (1623), the Reverend Samuel Purchas voiced the eye-for-an-eye attitude when he described both the slaughter of the Roanoke settlers in 1607 and the killings of 1622 by the Powhatans and Chickahominies in words as blood-curdling as any of those in the Old Testament: "Their murdered carcasses . . . speak, proclaim and cry, This our Earth is truly English."[88] (Opechancanough and his people no doubt felt in 1616 that their earth was truly Indian.)

The recovery of the Virginia settlers was as prompt and remarkable as the attack itself. Stoutly refusing to remove to the Eastern Shore for greater safety, the inhabitants set about repairing James City. Their indomitable confidence in this English enterprise was based on the firm belief that, as Bishop John Aylmer had proclaimed in 1558, "God is English." In 1623 in like vein, a "Gentleman of Virginia" wrote a ballad, *Good Newes from Virginia*, about the "savages" that "god . . . hath put into our hands, who English did abuse." This piece of doggerel took the literary form most popular in England and was hawked up and down the streets of London. It is probably the earliest piece of American verse.[89]

> So to *Opachankenowes* house,
>   they marched with all speed:
> Great generall of the savages,
>   and rules in's Brothers steed.
> But contrary to each mans hopes,
>   the foe away was fled:
> Leaving both Land and corne to us,
>   which stood us in stead.

Elsewhere, an English captain proceeded "with honor":

> Who comming took not all their corne,
>   but likewise tooke their King:
> And unto *James* his Citty he,
>   did these rich trophies bring.
> And divers ships are still abroad,
>   with hundreds for to find:
> Both corne and victaile from these foes,
>   that us'd us thus unkind.

As for those natives who "love the English fervently":

> We use them as we use our selves,
>   with selfe same curtesie.

> Great and most gratious mighty God,
> thy name be ever praised:
> Which late dids't bring thy servants low,
> who now they selfe hast raised.

Accepting the view that the best Indians were dead ones, the Governor and Council, at the end of August 1622, declared war and fell upon the aborigines wherever and whenever they encountered them. Before the year was out the Virginia Council reported to the Company that they had "anticipated" their desires by "setting upon the Indians in all places." The natives admitted that the English had slain more of them this year than had been slain since the beginning of the colony.[90]

The war for racial revenge was pursued vigorously during 1623 and 1624 by the English. They attempted by forced trade or in other ways to make the enemy give up such provisions as they had to the needy inhabitants at Jamestown. Failing in this, they continued to carry on the scorched-earth warfare, which had been instigated by the officials of the Company in London after the "Massacre." Systematically the colonists began to burn the houses and canoes of the Indians, break their fish weirs, and destroy corn, peas, and beans in the fields in order to starve them out. This practice produced immediate results despite the "small and sicklie forces" of the settlers. Those natives whom they could not drive out of the Peninsula between the James and the York rivers, they persuaded by diabolical blandishments to plant around their villages, and then the English destroyed the crops so late in the season that new ones could not be started. In all probability, more Indians died of starvation than by all other means combined.[91]

The duplicity and role of Opechancanough in staging the attack immediately became apparent. As a part of the strategy in the conflict, the London Company moved to deprive the natives of his leadership, promising that "if any can take Opechancanough himself he shall have a great and singular reward from us." In spite of many rumors to the contrary, the chieftain foiled every effort to capture him; he was forced, however, to become a fugitive, even from his own people.[92]

During the absence of Opechancanough, his deposed brother Itopatin endeavored to regain power and become once more the

Powhatan. In May 1623 the lame chieftain sent word to Governor Wyatt that if he would send ten or twelve Englishmen to Pamunkey to guarantee that his people might plant their corn in security, "he would deliver all of the Captive English he had and would also deliver his Brother Opechankano, who was author of the Massacre, into the hands of the English either alive or dead"—nothing came of this faithlessness, however, for within a few days the fugitive was at Potomac ready to represent all of his tribes.[93]

With grave concern, Governor Wyatt had written to England in April 1623 about the relations of "our people" with the natives, asserting that "without doubt either wee must drive them, or they us, out of the country, for at one time or other they play us false." Seemingly in contradiction to this opinion, the governor sent Captain William Tucker with twelve men up the Potomac on May 22 in reply to Itopatin's request to recover and fetch away the detained Englishmen and, ostensibly, "to conclude peace with the great Kinge Apochanzion."[94]

In a letter to his brother in London, Robert Bennett graphically described the lurid happenings at this meeting at Potomac in which the perfidy of the white men exceeded that of the Yeardley massacre of the Chickahominies in 1616. After "manye fained speches the pease was to be concluded in a helthe or tooe in sacke which was sente of porpose in the butte with Captain Tucker to poysen them." Both the officer and his interpreter had to taste the white wine before the wary Opechancanough would drink any, but it was "not of the same" as that given to the natives, for the King of the Potomacs and "some two hundred" of the Indians were poisoned; later the English killed "som 50 more" and "brought home part of their heads." Two other kings died, "and manye alsoe" for certain, when the English came back and fired a volley into the crowd, but somehow Opechancanough escaped unharmed. A little more than a year later the Earl of Warwick stated categorically that the unsavory Dr. John Pott of the local council and later acting governor had prepared the wine. "Soe this beinge done," Bennett concluded, "it wilbe a great desmayinge to the bloody infedelles . . . God send us vyctrie, as we macke noe question, god asistinge." At least one of the settlers, Peter Arundell, questioned the pious English phraseology about the natives: "Wee ourselves have taught them how to bee trecherous by our false dealing. . . ."[95]

By November 12, 1623, the Company in London had learned from letters that the fatal sickness, which carried away more settlers than the "Massacre," was no longer prevalent, crops were plentiful again, and Opechancanough and some 150 of "his great men" were slain—a statement that proved to be false. The confidence of the English soared, and by March 2, 1624, members of the Assembly were bragging to the Royal Commissioners that since the massacre of the previous year, not a white man had been lost in any expedition sent against the natives. Three days later orders were issued that beginning in July the colonists of each corporation were to attack the "Salvages" near by "as wee did last yeare," cut down or burn the corn planted by them "uppon hope of a fraudulent peace, with intent to provide them selves, for a future warr, and to sustain their Confederates." The councilors reported to the Company that "wee hold nothing injuste, that may tend to their ruine (except breach of faith)."[96]

In the course of advising the royal authorities about conditions in Virginia in 1624, Captain John Harvey laid out the relations between the colonists and the Indians: "They are ingaged in a mortall warre and fleshed in each others bloud, of which the Causes have been the late massacre on the Salvages parte, and on the parte of the Englishe a later attempt of poysoninge Opochancano and others." The great native leader had disappeared into the forest, and the lame Itopatin again had assumed control of the Powhatan Empire, determined to fight it out with the enemy.[97]

The Indians fought on desperately. In December the Council wrote triumphantly to the Earl of Southampton: "It hath pleased God this yeere [sometime after June] to give us a great Victorie over Otiotan [Itopatin] and the Pamunkeys, with their Confederates." The English numbered a mere sixty men, of whom twenty-four were kept busy cutting down corn—it was estimated that the corn destroyed would have fed four thousand men for a year. In this battle, Governor Wyatt learned "what the Indyans could doe, having mantayned the fighte two days together, and much thereof in open fielde." None of the English were killed and only fourteen were wounded by native arrows. "The Indyans were never knowne to shew soe great resolution," in what was the greatest stand-up fight between the two races before the Battle of Point Pleasant in 1774.[98]

Posterity has been no less mystified about the doings of Opechancanough after the attempt to poison him than were his enemies in Jamestown and London. Had he been present at the Battle of Pamunkey, it is likely that he, not Itopatin, would have been in command, and the whites would have observed that fact and reported it. "As for the natives," Captain William Pierce stated in his "Relation" (*ca.* 1628): "Sawsapen [Itopatin] is the chief, over all those people inhabiting upon the rivers next to us, who hath been the prime mover of them, that since the massacre have made war upon us. But nowe this last Somer [August 12] by his great importunity for himselfe, and the neighboring Indians he had obtained a truce for the present . . . being forced to seeke it by our continuall incursions upon him, and them by yearly cutting downe, and spoiling their corne."[99]

When the natives failed to abide by the treaty—stealing tools and hogs, and killing isolated colonists—the Governor and Council, after a year, resolved to terminate the peace. By their own admission, from the time it was concluded, the English had been awaiting "a fit opportunity to break it." After notifying "the great Kinge" [Itopatin] officially, the governor ordered the commanders at all plantations to go out against the aborigines and "utterly destroy them." The Council drew up plans for regular forays every November, January, and March to cut or burn Indian crops and fire their houses. It is easy to understand why the neighboring tribesmen were labeled by the Assembly as still being "our irreconcilable enemies." Nevertheless, in 1632, Governor John Harvey made peace once more with the Pamunkeys and Chickahominies but warned the colonists not to trust them.[100]

For two decades after 1624 racial difficulties were seemingly resolved. Itopatin faded from view, and one searches the records in vain for any further trace of Opechancanough; all that can be said is that he became a centenarian in 1644. Then suddenly, in 1644, he reappeared for one brief moment before he departed this earth like the final burst of a sky rocket.

The aged chieftain learned from some of the settlers that they were uneasy because "all was under the sword in England" (a fight had just taken place in the James River between a parliamentary ship and a royalist one). Rousing himself for one last mighty effort, he attempted to save his tribesmen. "Now was his time or never, to roote out all the English." Here too the facts are few, but that after twenty years, the Indian warriors would accept his leadership and keep his plans secret are most remarkable facts.[101]

In a general massacre (*his fourth*) on April 18, 1644, Opechan-canough and his braves surprised and slew "near five hundred Christians," chiefly on the south side of the James and at the heads of other rivers. A fitting tribute to the "great King's" organizing capacity was the exaggerated statement by one of his followers that all of the Algonkians within six hundred miles were conjoined to root out all of the white strangers in Virginia. Also the Indians gave the same explanation of this uprising that was applied to the "Massacre" of 1622: "They did it because they saw the English took up all their lands and would drive them out of the country." On their side, the English could only damn the natives for "their treacherous manner" of operating: "they have not the courage to doe it otherwise."[102]

"Opechancanough, by his great age, and the fatigues of war . . . was now grown so decrepit that he was not able to walk alone," Robert Beverley tells us, "but was carried about by his men, whenever he had a mind to move. His flesh was all macerated, his sinews slackened, and his eye lids became so heavy, that he could not see, but as they were lifted up by his servants." Such indeed was his condition when Governor Berkeley had him captured and brought a prisoner to Jamestown. Sir William treated the "ancient Prince" with respect and care, "hoping to get a reputation" by sending a royal captive to England who could raise ten times more fighting men than his enemies and also demonstrate the health and long life of the native Virginians.[103]

Opechancanough had been in Jamestown but a fortnight when one of the soldiers "basely shot him through the back . . . of which wound he died." In classic Indian fashion he had shown not the least sign of dejection at his confinement or wounds. The expected war of revenge took place, and the white men won. In October 1646, the governor made peace with the chieftain's successor, Necotowance. Among other things this treaty provided that if any native should venture within the limits of the English settlements, "it shall be lawful to kill him presently," but the white colonists might pass freely and safely "where they please through his dominions."[104]

Thus did the intelligent and tireless efforts of the great native American, Don Luis/Opechancanough, almost unremembered in our time, all come to naught and pass into the history of a troubled period. The prophesied third invasion had triumphed completely. And Jamestown was the scene of the final act of the First Lost Cause.

AFTERWORD

"Truth can never be proved," Lin Yutang wisely wrote, "it can only be hinted at." I do not contend in the present essay that I have *proved* that Don Luis and Opechancanough were one and the same person, but such conjectures as I have made are consistent with reason and common sense, and all of the known facts about this great Algonkian leader fit this theory, which clarifies much that before has been obscure. "Rule 1 in any serious reappraisal of puzzling historical facts," Joseph Alsop points out, "is the rule that forbids the flat assumption that what can have happened nevertheless did not happen."[105]

No one can be more fully aware than I am of the incompleteness of the evidence about this Indian. It is possible that further information may come to light that will amplify or alter my conclusions. But certainly now is the time to tell what we know of the life of a truly remarkable American. With such facts as we have he comes alive. He is too vivid a personality to remain concealed from American history.

II

The Old and New Societies of the Delaware Valley in the Seventeenth Century *

"This country of New England from Virginia southward, to the french northward at Penobscott is about 6 or 700 miles along the sea coast. Toward the South at a place called Delaware bay live some Swedes, about an hundred, and likewise some few Hollanders, which hinder the English from planting there, though some 20 familyes from Mr. Davenports plantation [at New Haven] have attempted to settle there. This river is a very great river, very fruitfull, and will contayne more people than all New England beside. I suppose this place for health and wealth the best place the English can set there [sic] foot in. If any leave the Kingdom I pray counsell them to this place, and here many will joyne with them who have seen the place."

This reliable account of the English settlements along the Atlantic coast, written in Boston on December 24, 1645, "when the weather was so cold that the Ink and pen freeze extreamly," was drawn up by Dr. Robert Child, "the Remonstrant," to fulfill a promise made to his friend Samuel Hartlib, the celebrated authority on English husbandry.[1] We are surprised to learn that the valley of the Delaware was so well known at Massachusetts Bay as early as 1645 and pleased that this accurate description of this still

* This essay is reprinted from the *Pennsylvania Magazine of History and Biography*, Volume C, Number 2, April 1976, pp. 143–72, with the permission of the editor.

unoccupied region was made by the colonist who undoubtedly was the person best informed about agriculture in North America. Understandably, Dr. Child was not aware of the fact that the region had been the home of a centuries-old native society that was now doomed to disintegration and collapse.

To understand the societies, the old and the new, one must view the area *as a whole*.[2] From 1600 to 1700 the entire system of waterways and the land drained by it formed a single geographical unit of greater historical importance than the four political divisions that comprise it taken together—Cecil County in Maryland, the Lower Counties on the Delaware, Pennsylvania, and West New Jersey. This was true before the advent of the Europeans and, to an unrecognized extent, it remains so in our own time.

The broad, unusually clear, and placid river served as an avenue giving access to, as well as unifying, the peoples who lived along its banks from the Delaware capes to the falls at Trenton, and also those situated up the many large "cricks" that were navigable for small craft for one to ten miles inland. The terrain along the east shore in West New Jersey was flat, and its soil, though fertile, was generally sandy; a mile or two in from the rich low-lying west bank of the river, the land of Bucks, Philadelphia, Chester, and New Castle counties was rolling and then hilly; and there too the soil proved to be productive under careful cultivation. Everywhere springs abounded, and the region was well watered.

1610–1638

In common with all history, a correct distribution of events through time is antecedent to a proper understanding of the Delaware Valley during the seventeenth century. Modern books dealing with the area often give the impression that the history of the Valley properly begins with the landing of William Penn and his followers. They ignore the "Original People," a loosely confederated group of tribes whose members spoke Algonkian dialects.[3] These Indians, the Lenni Lenape or Delawares, had been living in the Valley for hundreds of years when the century opened, principally on the west bank of the great river from its mouth to the falls.

In 1600 the society of the Lenni Lenape numbered somewhere between two and three thousand people who were flourishing, virtually unaffected by any contact with either the things or the people of Europe. They were not wandering hunters; they led a settled existence in small open villages, basically a life of peace and harmony. The average community consisted of five or six "longhouses" sheathed with strips of bark and each having a peaked or ridge roof. These early American multi-family dwellings varied from thirty to one hundred feet in length and from twenty to twenty-five in width, each of them sheltering several families. Contrary to white mythology, the natives kept their persons clean; youths swam daily in most seasons, and adults visited their "sweat lodges" weekly or oftener to take steam baths. In the open fields surrounding the longhouses, the natives grew tobacco and the "three sisters"—corn, beans, and squash. They also traded with distant tribes.[4]

The first meetings between the Indians of the Valley and Europeans occurred about 1610, but chaffering for pelts did not become common until after 1620 when, with Dutch thoroughness, skippers and traders from Manhattan began exploring the river and its tributaries. Surviving evidence indicates that it was the European *things* offered by the Dutch in exchange for pelts that initiated the transformation of the Indian society.

The Dutch made no effort in the early years to establish permanent settlements in the Valley; Fort Nassau (1626) and Fort Beversreede (1648) were merely trading posts maintained by a few men—theirs was a male society. At first beads and similar baubles attracted the Lenni Lenape, but it was not long before the Indians were insisting that, in exchange for their furs, the traders should give them firearms, ammunition, hatchets, kettles, hoes, and a variety of metal utensils; later on they asked for red and white cloth. The Hollanders also introduced the natives to brandy and, after mid-century, to rum. As the months and years passed, the Indians became more and more dependent upon European goods for performing all the tasks allied with daily living: agriculture, hunting, and household chores. Correspondingly, pottery-making and other native crafts and skills deteriorated from disuse. Gradually it came about that the peltry traffic was leading to unforeseen changes in the life of the Lenni Lenape.

With the white men's ever-mounting, insatiable demand for

beaver skins and other pelts, the Indian men gave up their age-old ways of life and became hunters and trappers exclusively, which meant that they were away from their families and fields for long periods of time. Furthermore they abandoned their instinctive habit of solicitously conserving wild life, especially the beaver and deer, and proceeded to slaughter beasts of all kinds indiscriminately merely for their skins. Later they killed them to provide meat for the white men.

The increasing demand for pelts stirred up competition among the Indians that quickly became a divisive factor in intertribal relations and inevitably brought hitherto friendly tribes into fratricidal conflict. Less numerous than the Lenni Lenape, the Iroquoian-speaking "White" and "Black" Minquas, or Susquehannocks, who lived in the Susquehanna River basin were more given to hunting than to aboriginal agriculture. They were active traders, procuring many of their skins from tribes to the westward, but by 1624, at least, they were encroaching on the territory of the defenseless Lenni Lenape, who greatly feared them. The Minquas congregated in villages surrounded by palisades that protected them from retaliation by enemies. When David Pietersen De Vries first encountered them near deserted Fort Nassau in 1633, it was clear that this warlike Iroquoian tribe already dominated the Lenni Lenape.[5]

Despite being technologically outmoded by the white newcomers, the Indians made great and lasting contributions to ·the development of the white society. Long before the coming of the first Europeans, the Minquas and, to some extent, the Lenni Lenape had worked out routes of overland travel, often no more than fifteen inches wide, along forest paths from the head of Chesapeake Bay to the Delaware River, and from its falls across New Jersey to the Raritan River. From their palisaded strongholds on the west bank of the Susquehanna, the Minquas would sometimes float downstream in their unsinkable dugout canoes, which were loaded with furs, to the Chesapeake Bay and then paddle up the Elk River to one of the portages leading to the Christina (Minqua Kill) or Appoquinimink creeks, or farther overland to the tributaries of the Schuylkill. These waterways and forest paths and portages constituted a series of inland travel routes that were to prove invaluable to the white traders.[6]

By 1638 the Indian society, whether that of the sedentary,

peaceful Lenni Lenape or that of the fierce, hunting, and trading Minquas, had been irrevocably disrupted and undermined by
the fur trade, alcohol, and *things* made in Europe. The final collapse
of the society as it existed prior to 1600 was ensured by the arrival
of numbers of white men who came with the intention of settling
permanently in the Valley. It ill becomes us of the twentieth century
to dismiss the process as the normal outcome of the clash of an advanced society with a backward one. Which was the superior,
which the inferior one? Of this we can be certain, however, the
seventeenth-century history of the Delaware Valley is, from the
Indians' standpoint, one of a grim disaster in which the ignorance
and greed of the victorious whites figured far more prominently
than their vaunted humanity.

1638–1675

The arrival of Peter Minuit with colonists in the spring of 1638
started a profound shift in the composition of the population of the
Delaware Valley. In addition to the few Dutch and English
traders, who do not exactly qualify as settlers, two new national
groups, Swedes and Finns, came in 1638 to live permanently in
the valley of the Lenni Lenape. The addition of four European
languages to the Algonkian and Iroquoian tongues complicated
communications for the inhabitants, and the presence of Calvinist,
Lutheran, and Congregational Christians confused the pagan majority who revered the Great Spirit. In 1655 the Dutch conquered
New Sweden, and in 1675, John Fenwick settled 150 English
Quakers on Salem Creek. If the sixty or seventy black slaves taken
from the Dutch by the English in 1664 were kept in the Valley,
they, together with a few French, added one more race and at
least two more languages to the already cosmopolitan character of
the area.

It was not the English, Irish, and Welsh Quakers led by William Penn, but the Swedes and the Dutch who first bought lands
from the Lenni Lenapi. Fort Christina, erected on the land purchased by Minuit at the mouth of the creek of that name, gave the
Swedes and Finns access to the peltry trade with the Minquas; it
also signaled the beginning of an enduring agricultural settlement

of members of the white race in the Valley. The Old World concept of absolute ownership of land, in opposition to the Indian idea of use only, and the fencing in of land, the visible sign of private property, was alien to Indian thinking. From the outset in 1638, the Swedes and Finns proceeded slowly and unobtrusively to develop communities of the European type. The Lenni Lenape seemed to be more than willing to "sell" land, that is, to allow the newcomers the use of it; and moreover they taught the Europeans the elements of aboriginal farming. In 1641, New Sweden was made up of tracts of land on the west bank of the Delaware between the bay and the falls, and on the east bank upstream from the bay as far as the mouth of Raccoon Creek. By the end of this period, these enclaves, as well as a few Dutch communities, provided the sole fixed element in an otherwise unstable situation.

The Scandinavians were never very numerous, and, though this might have been disadvantageous to New Sweden, the slow growth of population cushioned all contacts with the Lenni Lenape and the Minquas. These white men managed to get along very well with their Indian neighbors. In 1640 the settlers totaled a mere forty; and at the peak of development, in 1654, they numbered only 370 souls. Some of the Swedes and, later, a few Dutch learned to speak Algonkian and Iroquoian, which greatly facilitated trading, and the Indians welcomed the establishing of Swedish fur-trading posts up and down the river to tap the trade with the Minquas. This generosity may have stemmed from the Indians realizing what Governor Johan Printz did not learn until 1643–44, which was that the supply of beavers in their own valley had been exhausted.

The most important contribution of the colonists of New Sweden (half of whom were Finns) to the newly forming society of Europeans was family life. The Dutch mariners and traders had operated as a strictly male group until the 1650s, but the Scandinavians wanted wives and children and home life. "For one who wishes to remain here, he cannot be without a wife," Vice Governor John Papagoya wrote plaintively in 1644. "If one were in Sweden, there would be no want; but here one must himself cook and bake and himself do all the things women do, which I am not accustomed to do, and it is difficult for me."[7] Governor Printz had made a great point in 1647 and again in 1650 of the Company's sending over wives for the farmers. Three years later Printz contrasted the prospering farming families with his sorry lot of soldiers and company officials. William Penn, when he came to New Castle in 1682, particularly praised the sturdiness and civilized qualities of the colonists—Swedish and Finnish, as well as the Dutch who had come to settle after the conquest of New Sweden in 1655. He emphasized that they were "a People proper and strong of Body, so they have fine Children, and almost every house full; rare to find one of them without three or four Boys, and as many Girls; some six, seven and eight Sons: And I must do them that right, I see few Young men more sober and laborious."[8]

Coming from a forested land, the Swedes and Finns were skilled in the use of an ax to shape timbers, and upon their arrival in the Delaware Valley they erected weathertight houses or "cabins" of round-crossed or hewn logs, such as they had known in their homeland. Later they put up larger log structures—forts, Crane Hook Church (1667), and the blockhouse used for worship at Wicaco (1669). The few barns they built were also of cross-notched construction. The Czech Augustine Herrman imitated the Swedish method when he built "a Logg house Prison" at Bohemia Manor in Cecil County, Maryland, in 1669, as did the Quaker justices of the English court who, in 1681, ordered "a Convenient Logg house for a Prison" at Burlington. In general, as Jasper Danckaerts and Peter Sluyter noticed in 1679, in both West New Jersey and Pennsylvania the English built clapboard houses without regard to the superiority of the Scandinavian structures.[9]

Adaptability being the price of survival, the Swedish, Finnish, and Dutch husbandmen learned to imitate the tobacco and maize culture of the Lenni Lenape. They also shrewdly employed the natives to supply them with venison, turkeys, and other game until such time as they could raise their own domestic animals. Farms soon clustered around the Swedish trading posts at Christina, Upland, Tinicum Island, Passyunk, and, after 1658, when the Dutch sent over five hundred colonists of several nationalities, at New Amstel (formerly Fort Casimir and after 1664 New Castle). Most of the colonists, whose numbers had doubled by these late arrivals, lived on the west bank of the Delaware between New Amstel and the falls; "a handful" of Swedes, Finns, and Dutch resided on the east bank; they planted orchards and grew wheat, rye, and barley. From Virginia and Manhattan they imported cattle, which multiplied "greatly with great wonder" by mid-century. Jasper Danckaerts marveled in 1679 at the sight of an ox on Tinicum Island as large as those of Denmark and Friesland.[10]

As late as 1685, Francis Daniel Pastorius believed that "the old inhabitants" were poor agriculturalists; "many of them have neither barns nor stables, let their grain lie unthreshed for several years under the open sky, and their cattle, cows, swine, etc., run summer and winter in the underbrush, whereby they gain but little benefit from them." What this superior husbandman was looking askance at was pioneer agriculture, and he failed to see that these farming colonists were producing enough grain and cattle by the close of this period to enable them to feed many of the advance contingent of the Quaker invasion.[11]

From 1638 to the coming of William Penn and his followers, tobacco dominated agriculture and trade. After becoming familiar with the paths and portages the fur traders had learned from the Indians, the colonists gradually converted them into thoroughfares—cart and "rolling roads"—that enabled tobacco grown in Maryland and Virginia to be brought overland to Fort Christina and New Castle. Governor Printz shipped 20,467 pounds of the leaf to Sweden in 1644, of which about 5,000 pounds were produced in New Sweden and the balance in Accomac, Virginia. A decade later, the governor made a contract with Edmund Scarborough of Accomac, who often traded in New Sweden, to supply 80,000 pounds of the "Stinkingeweede" of America to the homeland. The traffic swelled continuously and, down to 1700, this "chopping herbe of Hell" continued to be the principal export from the lower Delaware Valley.[12]

As the value of tobacco exports mounted, the fur trade declined precipitously. The latter had been very profitable in the early decades of settlement and probably reached its peak in 1649 when the white men traded for between 7,500 and 10,000 beaver pelts. After that year the Dutch, who had more goods to barter with than the Swedes, took over most of the trade in peltry. One of their reasons for seizing New Sweden in 1655 was that "thousands of Beavers can be bought here [on the river] and around the Schuylkill or [Fort] Bevers reede" from the Black and White Minquas.

When the Minquas were defeated by their Iroquois kinsmen in 1675, the traffic in furs, something the new English rulers counted on, dwindled to almost nothing.[13]

The slow extension of settlement by the white men was an important feature of the economic transformation of the Delaware Valley, for it was accompanied by the spreading of the culture of tobacco and grain, as well as cattle grazing. The improvement in the routes of travel was a prime factor in this expansion. As early as 1644 Governor Printz had expressed the desire to drive the Lenni Lenape out of the Valley: "Then one would have a passage free from here [New Sweden] unto Manathans, across the country, beginning at Zachikans [the falls]." As noted earlier, the Dutch had explored the river and its tributaries, and in 1654, Peter Lindeström, a Swedish engineer, recorded the waterways on his excellent "Map of New Sweden."[14]

For commercial, as well as political reasons, the Dutch, who had become a decisive factor in the development of the Valley during the fifties, sought very early to open an overland route from their base on Manhattan to the Delaware River that would be safer than the all-water one. From the Raritan River in New Jersey, two Indian paths, known as the Upper and Lower Assunpink trails, crossed the level country to the falls of the Delaware. In order to surprise Printz and the Swedes in 1651, Director Peter Stuyvesant "came here himself overland, accompanied by 120 men; but eleven vessels came by sea, which met him here at Fort Nassau [Gloucester]."[15] This long and successful march along the Upper Assunpink Trail to the falls of the Delaware by the largest military force yet raised in the colonies was a remarkable feat and, considering the conditions along the way, a triumph of logistics. Thenceforward the two New Jersey routes were in continuous use.

Governor Rising told the Swedish West India Company in 1654: "Hereafter it would be well worth while to settle Christina Kill, in order that one might be more secure against Virginia [actually Maryland], and besides to carry on a trade with them, making a passage [canal] from their [Elk] river into the said kill, by which we could bring the Virginia goods here and store them, and load our ships with them for a return cargo." This earliest proposal for a Delaware and Chesapeake Canal had been made to Governor Rising by Thomas Ringgold of Kent County, Maryland. The governor hoped to purchase a tract of land from the Minquas at the present Elkton, and "then this could be well brought about, and we could carry on the best trade with them [the Minquas] there." He also stated that the shortest portage is from the Appoquinimink Creek to Head of Elk, and that the English in Maryland "are now beginning some trade from their own river in that direction."[16] A year later, Peter Stuyvesant defeated the Swedes and New Amstel became the principal trading port on the Delaware.

Some of the colonists of New Sweden—servants of the trading companies who desired to avoid harsh treatment, and freemen who wanted to acquire cheap land—cleared out prior to the change over in government and located around the head of Chesapeake Bay (one piece of land in Maryland became known as "none so good in Finland"). In a complaint about the tendency to flee inland, Governor Johan Rising stated that "the English draw our people to themselves over to Virginia (Saverne [Severn]) as much as they are able and keep those who deserted thither last year. They largely ruin our trade with the Minques." The draining off of settlers continued; in 1676 the Duke of York's Laws prohibited all persons in the middle colonies from traveling without a passport, and the first laws promulgated in Pennsylvania (1682) required "Unknown persons" to carry passes for each county. In the neighborhood of Bohemia Manor in 1679, Jasper Danckaerts, who carried a passport, learned that "They are almost all Dutch who live here."[17]

After the conquest of New Sweden in 1655, many Dutchmen removed to Maryland and settled around the head of Chesapeake Bay.[18] Among them was Augustine Herrman, the most important individual connected with the Delaware Valley in the two decades before the arrival of William Penn in 1682. Born in Prague in 1605, he grew up in Holland. When he arrived in America he had a knowledge of the Czech, Dutch, French, English, and German languages, an ideal equipment for the roles he was to play: trader, surveyor, diplomat, and principal founder of the fur and tobacco trades in this cosmopolitan valley.

Herrman made his first public appearance at the treaty with the Lenni Lenape of Passyunk at Fort Nassau in 1648, but it was a diplomatic mission to Annapolis undertaken for Stuyvesant in 1659, which took him overland from New Amstel southward to the river Elk, that changed the course of his life. Mightily attracted by the rich land he observed around the Chesapeake area, he moved to Maryland a year later. In 1663 he and his family were naturalized; he received a grant from Lord Baltimore of 13,000 acres in Cecil County, where he erected Bohemia Manor. He had been a denizen of Maryland a very brief time when, in 1661, he began planning a cart road from his estate to New Amstel. Ten years later, two of his sons settled on the Delaware at Augustine Manor, a second tract belonging to their father. And in this same year, Herrman's dream of the road came true when the Governor and Executive Council of New York agreed to clear half of "a way" from New Castle to Bohemia Manor, and Augustine Herrman and the Marylanders agreed to open the other half. Shortly, along this Bohemia-Appoquinimink route—"a large broad wagon road"— carts could be seen carrying tobacco to the ports of the Delaware, and people were traveling on foot or on horseback in numbers sufficient to warrant requiring them to carry passes.[19]

After ten years of painstaking surveys, Augustine Herrman prepared a map of *Virginia and Maryland as it is Planted and Inhabited This Present Year of 1670*, which elicited from Lord Baltimore the tribute of being "the best mapp that was ever drawn of any Country whatsoever." (Plate 3) Engraved and published in London by William Faithorne in 1673, this large and beautiful draft was the most important cartographical endeavor produced thus far in the colonies. Although Herrman's map depicted the Delaware Valley only as far as Matinicunk Island (Burlington), he did indicate the path overland to Navesink and New York; and the rendition of the Bohemia Manor-New Castle streams was indeed revealing.[20]

The true significance of the intercolonial routes of travel in the Valley became evident in May 1672, when George Fox and three other Friends, with the help of two Indian guides, rode "towards New England by land" from Miles River in Talbot County, Maryland. Proceeding by way of Bohemia River and from there to New Castle, they crossed the Delaware River in a sloop and again rode through woods and along the Burlington Path to Middletown in East New Jersey—"and were very glad when we got to a highway." There, another Friend took them in his boat on a whole day's passage to Gravesend on Long Island. After spending six

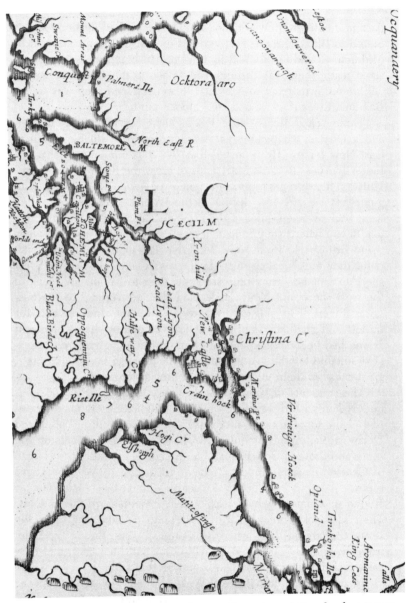

A section of Augustine Herrman's map, *Virginia and Maryland as it was Planted and Inhabited this present Year 1670*

days in meetings at Oyster Bay, the Quakers sailed up Long Island Sound to Rhode Island. In August, Fox and his party returned to Middletown and in the next month, guided by hired Indians, they again made their way through the woods to the Delaware River, and then downstream to New Castle and retraced their way to the Bohemia River route and on to Talbot County.[21]

Three years after George Fox's journey, William Edmundson and a Quaker companion sailed from New York to Shrewsbury, East New Jersey. He reported that "we took our journey through the Wilderness towards Maryland, to cross the River at Delaware-Falls." Their progress fell short of expectations, for the Indian guide lost his way and "left us in the Woods." The next day they went back over ten miles of the previous day's journey to the Raritan River, where they found "a small path" at the landing place (New Brunswick) from New York that led to the falls, at which point they took a canoe over the Delaware and followed the west shore to Upland. They recrossed the river to Salem, and back to the west shore and New Castle, and thence overland to the Sassafras River. The old routes traversed by the Indians and fur traders and gradually improved and developed by the Swedes and Dutch had become avenues for communication, travel, and transport suitable for the age and place. The English had but to make greater use of them in the linking of their colonies.[22]

The conquest of New Netherland by the English in 1664 was undertaken with the twofold object of eliminating Dutch competition in the tobacco trade and consolidating the Hudson Valley and Delaware Valley settlements with the Chesapeake colonies on the south and those of New England to the north, all under the control of Charles II. The English traders, according to William Tom's report of 1671, neglected to learn either the Algonkian or Iroquoian tongues, which was bitterly resented by the Indians. Their sachems were actually threatening to make war upon the settlers, charging that "where the English come they drive them from their lands," as in northern Virginia and Maryland, and they were fearful that a similar fate would be theirs.[23]

It was perhaps inevitable that as the society of the Europeans took permanent form, that of the red men should visibly decline. The truly peaceful relations of the white men with the Indians occurred during the Dutch-Swedish era down to 1675, for there were never any wars between the Indians and these predecessors of the Society of Friends, such as vexed the inhabitants of the Hudson River Valley. The Swedes, the Finns, and the Dutch had sought to treat the Indians well. They were always careful to purchase land before they occupied it, and the Swedes charged the Minquas lower prices than the Hollanders did for trading goods; all three learned both the Iroquoian and Algonkian dialects and taught the naked Indian to wear European cloth.[24] In the same way the white men tolerated each other's faiths—Calvinist and Lutheran forms of Christianity—they refrained from interfering with aboriginal worship of the Great Spirit. In these, and in so many other ways, they set precedents for William Penn and the Quakers.

During the period from 1638 to 1675, Europeans prepared the way for the coming in force of the Quakers under Penn. Ultimately, they dealt the *coup de grâce* to the native society by gradually substituting their own technology for that of the Indians by clearing the land, starting farms, fencing fields, building houses, establishing courts of law, and forming towns. Not only did they settle along the west bank of the river and reach out toward Maryland; they opened West New Jersey for occupation and built the towns of Salem and Burlington. They laid the foundation for the cosmopolitan society composed of many European nationalities, which has ever since been an outstanding feature of life in this region. And by introducing black slaves well before 1644, they added a third race in the Delaware Valley. In 1675, the inhabitants, black and white, may have numbered around seventeen or eighteen hundred.

Peter Alricks noted in 1664 a sharp drop in the Lenni Lenape population, estimating that there were not more than a thousand living on both sides of the Delaware. An early settler in West New Jersey, Robert Smythe, thought the Delawares there were tractable, and saw no need "to new-mold, displace, or remove, contend or quarrel" with them as long as their lands were purchased before settlement began. "And the *poor Creatures* are never the Worse, but much better, as Themselves do confess; being now supplied by the *English*, in the way of Truck and Trade, with whatsoever they want or stand in need of. And they Hunt and Fish, as here-to-fore, except in Enclosed or Planted Ground."

From the falls of the Delaware in 1680, Friend Mahlon Stacey remarked sanctimoniously, with the arrogance of an Englishman, that he believed God "*is determined of a small number to make a Great and Strong Nation . . .* and the Lord is mightily bringing it to pass, in His Removing the *Heathen* that know him not, and making room for a better People that fears his Name. 'Tis hardly credible to believe how the Indians are wasted in Two Years Time . . . and how the English are increased both in Cattle and Corn in a little Time."[25]

The earliest settlers turned over to the Quakers a well-explored region, and a system of roads, paths, portages, and waterways that made overland communications possible and, for that age, relatively easy. Moreover, they had linked the plantations of New England with those of the Chesapeake and the Carolinas, so that in the coming wars the colonists were able to present a strong front to the French. From both an intercolonial and an imperial point of view, the joining together of the twelve continental colonies under one common law, one language, and one flag was an achievement of incalculable significance.

In 1681 the valley of the Delaware was physically, economically, and socially ready for the arrival of the Quakers who, as a result, never had to undergo many of the trials or perform the difficult tasks of early pioneers. The dazzling spectacle of the Quaker success, 1681–1700 and beyond, should no longer obscure the very important contributions of the aboriginals and the first traders and colonists.

1675–1700

The most impressive and far-reaching development in the insular and continental possessions of the English during the second half of the seventeenth century was the emergence in less than two decades of a new and strikingly successful society in the valley of the Delaware. By 1700 the course of this predominantly English society had been fixed.

Many writers have examined the institutional and political histories of each of the provincial components of the society of the Valley, but primarily it has been the role of the Quakers that has attracted "official" chroniclers and apologists. A preoccupation, at times amounting to an obsession, with the contests for power and the torrid factionalism in Pennsylvania and the Lower Counties especially has resulted in an unfortunate neglect of the truly spectacular rise and growth that went on in the entire region, a phenomenon that, regardless of men and measures, proceeded at a pace hitherto unparalleled and encompassed all other activities. Although a fully rounded survey of the process in this essay is patently impossible, one may draw attention to several unrecognized and uninvestigated aspects and qualities of the white society of the Delaware Valley that, without a grasp of their peculiar nature, leave its history obscure.

It is incumbent upon the investigator to develop a deep sense of companionship with the men and women who built this society, and therefore we begin with the people. The proportions of the three races shifted with the introduction of large contingents of Europeans. The society and life of the aborigines—disrupted by the tribal competition in the fur trade, liquor, and disease—ended with the radical decline of the original population of 1638. Dutch and Swedish colonists agreed about this. Gabriel Thomas reported in 1698 that "the Indians themselves say that two of them die to every one Christian that comes here." The Minquas disappeared from the Susquehanna Valley after their total defeat by the Iroquois in 1675; and Penn's three land purchases of 1682, 1683, 1684, forced the Lenni Lenape to move westward to the headwaters of the Brandywine and, eventually to the Susquehanna Valley. By 1700 the Indians had virtually ceased to be a determining factor in the lives of the Valley people. Actually they never were a real threat to the European settlers of this era, and now the "civilization" of the white men had overwhelmed them.[26]

Some sixty or seventy members of the black race had been held by the Dutch, as has been noted, when the English took over in 1664. Whether or not those particular blacks remained in the Valley has not been ascertained, but we have the assertion of Herman Op de Graeff in 1685 that "We have Blacks or Moors here also as slaves to labor." The slaves were numerous enough by 1688 to arouse the concern of Francis Daniel Pastorius and other Germantown Quakers to the point that they issued their famous protest against holding Negroes in bondage. In spite of their remonstrance, however, the number of black slaves unquestionably increased from 1690 to the end of the century.[27]

With the arrival of the first permanent white settlers in the Valley, who proved immediately that they could use the skills of the aboriginals and support themselves in the New World, the eventual dominance of the white race was ensured. Immigration varied from year to year, of course, in numbers, nationalities, and choice of place in which to locate.[28]

The Swedes and the Dutch had been the first to settle in the Valley, and many of them stayed on after the English took over in 1664, but settlers from the British Isles predominated. The white residents numbered between fifteen and eighteen hundred in 1675, and in the years before 1682 additional English, Irish, and Scottish Quakers arrived in West New Jersey to take up lands on the liberal terms offered by the Proprietors. To serve as market centers for these people, Salem was established in 1675, and two years later Burlington was founded, for by that time the east bank of the Delaware was dotted with farm buildings on cleared land from Salem Creek up to the falls.[29] With the founding of Pennsylvania, however, immigration to West New Jersey slackened.

Even before the English acquired the settlements on the lower Delaware from the Dutch in 1667, it may be recalled, some of the Swedes, Finns, and Dutch had moved over to the area around the head of Chesapeake Bay. Many of them had been living along the west bank of the Delaware from Shackamaxon south to Appoquinimink Creek since 1650, and by 1675 were successful farmers. They raised cattle and wheat and grew enough tobacco to add substantially to the growing number of hogsheads of the leaf brought overland from Cecil County, Maryland, to make up shipments by water to Manhattan.[30]

The Quaker immigrants of West New Jersey made remarkable progress on their farms between 1675 and 1682. Unsurpassed orchards of apples, peaches, and cherries plus fields that yielded quantities of marketable wheat and Indian corn were the rule, and already Jerseymen were known for their excellent cider. By 1678 they were freighting a "great plenty of horses" from Burlington to Barbados and Jamaica. When the Quakers began to arrive by the thousands after 1682, the husbandmen of the first settlements in the Valley were ready to join the Lenni Lenape hunters in feeding them;[31] there was no starving time, nor even a protracted food shortage.

Land was cheaper on the west bank of the river, and consequently after 1681 more than a few Jersey Quakers moved across the Delaware to Upland (Chester) and to the good land at the falls in what would become Bucks County. Shortly thereafter thousands of English, Irish, Scottish, and Welsh Quakers, attracted by William Penn's *Account* and *Further Account* of the goodness of his land, flocked to the new refuge, followed in a brief time, but in smaller numbers, by Germans from the lower Rhine Valley, Hollanders, French Protestants, and some Danes. Presbyterian Scottish and Scotch-Irish servants began to arrive after 1690, and so it went, the influx of Europeans continuing unabated. About fifteen hundred persons were added each year from 1690 to 1700 to the thousands already in the Valley. At the close of the century one can estimate that it contained about 25,000 inhabitants of all races and many nationalities distributed somewhat as follows:[32]

West New Jersey	3,500
Pennsylvania and the Lower Counties	20,000
Cecil County, Maryland	1,600
	25,100

The arrival of all these new inhabitants intensified still further the cosmopolitan nature of the population. Linguistically, communication among the people became more difficult as two Indian and eight or nine languages or dialects could have been heard at the Philadelphia market: in Germantown, a knowledge of *Hochdeutsch* was necessary for conducting any kind of business. The Welsh, Swedes, and Germans formed linguistic as well as national enclaves among the English. This permanent mélange of peoples of 1700 adumbrated what much of the later United States would be like, and it was bound to affect both economic and social life and the politics of the Delaware Valley.

The central fact about the rapid peopling of the area in this period was that it was accomplished for the most part by *families*. The number of Quaker families was second only to the number of Puritan families that settled New England. In age, the Friends ranged from infants born at sea to very old men and women. The ratio of females to males was very close, and the population rose rapidly because of a high birth rate and a low death rate. Such conditions combined to speed the formation of a sound, English type of society; men, women, children, servants, and hired laborers.[33]

The English-speaking settlers from the British Isles outnumbered all of the other nationalities taken together, but not by as big a margin as was formerly supposed. In numbers, wealth, trade, and influence, the members of the Society of Friends overshadowed their fellow colonists. On the other hand, the Delaware society turned out to be far less homogeneous religiously and socially than is generally stated. Early in 1685, for instance, "about 700 persons" came to Pennsylvania "From Carolina (one of the finest places in the world)" to "live under that peaceful regime." These immigrants were certainly not Quakers.[34]

Of major importance to the strength of the society was the coming of individuals and groups of colonists from Rhode Island, Massachusetts Bay, New York, the Chesapeake colonies, and the West Indies—Barbados, Antigua, and Jamaica. These colonists, most of them Quakers, provided Pennsylvania and West New Jersey with a select personnel possessed of wide colonial experience, considerable wealth, useful intercolonial connections, and political insight that enabled them to figure prominently in the creation of the bourgeois aristocracy noted by travelers, such as Dr. Benjamin Bullivant, in the nineties.[35]

Land in small or large parcels was sought by the newcomers, who soon located themselves up and down the Delaware, Schuylkill, Appoquinimink, Christina, Rancocas, and other streams of the six counties of Sussex, Kent, New Castle, Chester, Philadelphia, and Bucks, and in West New Jersey up to the falls. Farms appeared as much as fifteen to twenty miles inland from the west bank of the Delaware, to Plymouth Meeting, for example. On the excellent map made by John Worlidge and published at London in 1700 by John Thornton, this expansion, except for the plantations of the Elk-New Castle region, is clearly indicated.[36] (Plate 4)

The first act of a new inhabitant, after purchasing or acquiring a tract of woodland or a lot in town, was to erect some kind of temporary shelter. Following that he and his family usually spent

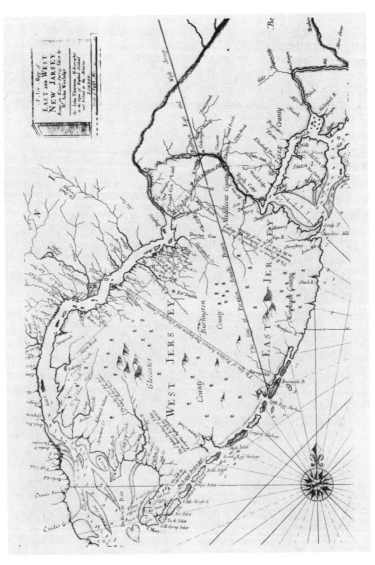

The Delaware Valley Settlements, 1690–1691

many long hours clearing the land, putting up fences, and then planting corn and wheat.

Except in the vicinity of New Castle and on farms in Cecil County, Maryland, few of the Swedish-Finnish log houses were erected, though time had proved they were the easiest to build and the tightest pioneer shelters. An exception was Jeremy Collett of Chester County who did have "a logg house" in 1686. It is a truism that colonists are always conservative, and the recently arrived Quakers were no exception; furthermore few of them were skilled in the use of an ax. The "clabbord" framed houses they built were invariably drafty.[37] In rural districts almost all of the household furniture was fashioned by the farmers themselves: homemade chairs, benches, stools, tables, cupboards, chests, and beds of walnut, oak, cherry wood. When the owners prospered, their second dwellings were increasingly constructed of brick or the lovely Pennsylvania field stone.

Once a farm was a going concern, the farmer would usually rebuild his first shelter, add to it, or erect another house, and it is reasonable to suppose that the Quakers who flooded into the Valley in 1682 would have needed about three thousand dwellings. Inasmuch as about three-fourths of the settlers were farmers, they also needed barns, "cattle houses," outbuildings, and mills, as well as warehouses, and public and religious edifices. Building for permanence proceeded briskly after 1685 in both town and country. In 1690 John Goodson wrote that "Building . . . goeth on to Admiration." The labor of many craftsmen must have been essential to construct the five thousand or more structures needed by the 25,000 inhabitants living in the Valley at the end of the century. What with building and rebuilding, the providing of shelter and housing for both men and beasts became the principal local industry, particularly in the last eighteen years.[38]

In the growing town of Philadelphia, and also Burlington, Germantown, Chester, and New Castle, housing was always a pressing problem. Building materials arrived by cart or in small boats from some distance away. Reports had it that in the three years 1682–85, six hundred houses (good and bad) went up in Philadelphia. After much rebuilding, Dr. Bullivant in 1697 thought it "generally very pretty"; he deemed some of the houses "large and stately dwellings," which "do already shew a very magnificent

City." "Ordinarily," the Bostonian concluded, "theyr houses exceeded not our second rate buildings in London," but did exceed "most shire towns in England."[39] This widely traveled Anglican gentleman rated the seaport on a par with New York, Newport, and Boston only fifteen years after its founding.

A lucrative lumber industry developed out of the demands from all sides for building materials—heavy timbers for framing, boards of many sizes, clapboards, and shingles, plus pipestaves, barrel heads, and hoops for coopering and export, ship's timbers, and thousands of cords of firewood for Philadelphia and the smaller towns as well as for consumption on the farms. Lumbering was, in itself, one aspect of clearing the land for farms. The preparing of many square miles of thinly forested land (as compared with the dense stands of timber in the Chesapeake colonies and New England) brought about a rapid and frequently wasteful destruction of the woodlands in the Valley, an immense ecological loss unnoticed at the time.[40] Construction also stimulated the rise of allied industries, such as brick making, stone quarrying, and lime burning, all of which required increasing numbers of carts and the building of roads.

Although the economic history of the Delaware Valley is yet to be written, several of its determining features during this period may be sketched briefly. It should be emphasized that the extension of areas under cultivation took place continuously, year by year, with the result that some farms were well developed and productive at the same time that others were only recently cleared and not yet yielding a subsistence for the farmer and his family. Agricultural development was progressive: first came the time of investment of capital and living on purchased provisions; next the state of subsistence farming; and ultimately the annual production of a surplus of grain, tobacco, and livestock for sale in the local markets, or for export to distant places. In these successive stages, the population of the entire Delaware Valley participated in a rough regional division of activities.

After a residence of eight years in Pennsylvania, William Rodeney declared in 1690 that "Husbandry is generally followed there, and is highly commended by our Neighbouring Provinces, Virginia and Maryland, for surpassing them." Besides the crops of corn and wheat grown for export, oats, barley, and rye produced fodder for horses and the principal ingredients for "the strongest beere in America," which was so much relished in Barbados and Jamaica. "The Country-men," Richard Morris stated at this time, "finding the Profit now coming in, do clear away the Woods, Plow and Improve their Lands," and raise not only grain but also hemp and flax for rural weavers of cloth; they also feed "great stocks" of oxen, cows, hogs, and "some Sheep" for their flesh and hides. Meat was cheap in local markets, and many beasts were butchered, salted down, and the meat shipped in barrels to the West Indies. Rural millers also sold quantities of flour to Philadelphia bakers who made bread and "bisket" for Caribbean slaves and ships' crews. "The Improvement in the Country of all Manner of Husbandry" stimulated commercial agriculture and the flourishing of coastal and overseas shipping, which, in turn, enriched many artisans and craftsmen and the merchants of the city of Philadelphia.[41]

As people moved inland away from navigable streams and farms appeared throughout the Valley, internal improvements (as a later age labeled them) were promptly made. To enable the farmers to carry their produce and drive stock to Burlington, Philadelphia, or New Castle, or to the mills and wharves near by, cart roads fanned out from such centers, and "Overseers of Highways," set up by the county courts, or the justices acting themselves, became responsible for their upkeep and improvement. "London Bridge" and "the Yorkshire Bridge" at Burlington were a part of this program, as was the structure spanning Crum or Poquessing Creek for the "King's Road," which ultimately ran up the west bank from New Castle to the falls of the Delaware. For the same reasons, that "traveling for men and beast may be more Easie, safe, and certain," ferries were instituted to eliminate the difficulties of crossing such wide streams as the Christina, Rancocas, and Neshaminy, as well as the Delaware and Schuylkill rivers.[42]

Communications by water were improved as small wharves were built at suitable landing places up the many creeks where sloops and wood boats, and a variety of other small craft could take in cargoes of country produce and firewood destined for Philadelphia. To accommodate the rivermen engaged in this essential activity, boat- and ship-building yards turned out a mounting number of wherries, shallops, sloops, and large ships fabricated chiefly from oak and other trees growing near by. By 1686 established Delaware Valley farmers were exporting the produce from their fields, pastures, and orchards to the West Indies, New England, New York, and the Chesapeake colonies. As David Lloyd announced that very year, the region had become "the Grainary of America."[43]

The cart roads, bridges, ferries, and river craft together served to unify the society of the Valley and, further, to speed communications up and down from West New Jersey to Maryland. Weekly, by 1695, a post rider traveled along a route extending from Portsmouth, New Hampshire, which passed through Burlington and Philadelphia, to New Castle. From Burlington, Dr. Bullivant set out for New York with the post rider and two Manhattan gentlemen in July 1697. He had just arrived from New Castle where he had been much impressed with the technological ingenuity of the settlers along the Appoquinimink Creek, "by which you may come to a neck of Land 12 myles over, Crosse which are drawn goods to and from Mary Land and Sloopes, also of 30 tunns, are carried over land and in this place on certain sleds drawn by Oxen, and launched again into the water on the other Side."[44]

Economic activities in the Delaware Valley were paying off in these last years of the century. In pounds sterling the total investment must have been large, for houses of brick and stone were not uncommon; nor should the capital invested in many warehouses, artisans' shops, wharves, bridges, and miscellaneous structures be overlooked. Buying and selling land for speculation forced values upward, and income from rents was substantial. Today it is impossible to make precise calculations, but the creation of wealth between 1682 and 1700 in improved farm land, rural buildings, urban houses, commercial and maritime facilities, together with income property, represented a profoundly important achievement. Circulating money, it is true, was scarce, but so it was in other colonies: this was a time of barter and book transactions. Provincial politics may have been confused and disorderly, but the soundness of the new society of the Delaware Valley was a triumph.

The society was financed not alone by the capital of Quakers in England and a small group of rich merchants but also by smaller sums invested by thousands of artisans and the great body of farmers along with their own and their families' labor. The total amount of money poured into "the Holy Experiment" over two decades must have exceeded the more than £400,000 that the English Puritans were estimated to have invested in the Massachusetts Bay colony, 1629–42.[45]

The remarkable religious and humanitarian accomplishments of the Quakers are well known, but their important secular, or worldly, achievement goes unheralded, even by nearly all of their own historians. The first of twenty-four ways by which the Friends supported activities of many kinds, listed by the apostate George Keith in 1702, was "their Established Weekly, Monthly, Quarterly, and yearly Meetings." There was also a variety of specially designated meetings: men's, women's, even children's; meetings for sufferings; meetings for the care of the poor; meetings for trade; and in 1700 a meeting "more particularly for the Negroes once a week" in Philadelphia. Some of these assemblages lasted several days or a week, at which time local Quakers offered "great hospitality to all friends and others." Through their quarterly and yearly meetings especially Friends succeeded in "keeping their Trade within themselves and maintaining a strict Correspondence and Intelligence over all parts where they are." Keith's twenty-second listing was that "By the People's great liberality to all their itinerant Preachers, putting many poor Mechanics, Servants, and Women that have no way of living . . . into such Ways of Trade and business, whereby to live plentifully, by which many of them combined farming with the exercise of their chosen mysteries."[46]

In planning and devising the superb organization of the Society of Friends for religious purposes, George Fox opened an intercolonial means of travel and communication that was promptly put to secular use. As George Keith observed, trade, as well as religious matters, was a part of the discussions in Friends' meetings; the Quakers were merely improving upon the late medieval custom of confining trade within the groups that were most trustworthy, such as the family, fellow countrymen, or merchants of a given region. The contribution made by this genius was the application of familiar organizing principles to the entire membership of his faith and the extension of its operation to a vast geographical area.

The New England Yearly Meeting, the first of its kind anywhere, had been convening at Newport, Rhode Island, for many years and was the model for the yearly meeting begun at Burlington in 1678, and which, after 1684, met alternately at Philadelphia and Burlington. This annual gathering, "essentially rural in its characteristics," embraced the Quakers of the whole Delaware Valley, Maryland, and parts of Virginia. By the end of the period, it included fifteen regular weekly meetings in West New Jersey and more than twenty-one in Pennsylvania, six of them in Philadelphia. "We have great and large Meetings in this Town and Province," John Goodson advised Friends in England in 1690. Both in scheduled sessions and perhaps even more informally, the Quakers who attended the various meetings discussed trading conditions, standards for goods and transactions, the care of the poor, exchanged shipping news, and worked out plans for the economic advantage of all Friends. Anglicans in the Chesapeake colonies, Congregationalists in New England, and Jews everywhere acted in much the same way, only their organizations could not compare with that bequeathed by George Fox to the Friends.[47]

No mercantile community anywhere could boast of as many individuals who possessed an equal familiarity with geography, topography, and maritime conditions, or who had gained experience all over the colonies, as had the Society of Friends. In 1672 it was the founder and his companions who first traced what might be called the *Quaker Overland Route* from Maryland to Rhode Island. In his tract of 1685 praising the *Good Order Established in Pennsylvania and New Jersey*, Thomas Budd pointed out that until local sheep could be bred and raised, all of the wool needed for household weaving might be imported from the Friends in Rhode Island and Providence Plantations.[48] Family connections and personal acquaintances throughout the colonies, in the Caribbean, in the British Isles, and even in some European countries were kept alive by oral reports and regular correspondence. By means of the discipline and assistance dispensed in the meetings, the energy, ambition, and worldly interests of the members were harnessed and directed along approved avenues to prosperity for Friends everywhere. Quaker merchants did not hesitate, however, to do business openly with honest men of other faiths—Anglicans, Lutherans, and Dutch Calvinists—who lived and labored in the Valley. Toleration and trade went hand-in-hand.

The new, burgeoning, and flourishing English society presided over by the Society of Friends had, by 1700, displaced "the Original People" in the valley of the Delaware. The Englishing of this rich region was, from the economic and social point of view, as profound an influence in American colonial history as Puritanism had been a half-century earlier, involving, as it eventually would, more people and a greater extent of territory. It determined that the English language, common law, commerce, crafts, industry, culture, and political organization would prevail permanently over the ancient Indian or other European influences.

III

Right New-England Men;
or,
The Adaptable Puritans[*]

The success or failure of any colonial enterprise known to history has always been determined ultimately by the kinds of responses made by the settlers to their new and strange environment. In that part of North America labeled "New-England" by Captain John Smith on his map of 1614, the immigrants from the Old World had to confront totally unfamiliar and unusual human elements and natural conditions. As months and years passed, through constant interaction between the planters and their surroundings, the "right N. E. man," as Edward Johnson characterized the malleable and industrious English settler in 1650, had already begun to reveal attitudes, habits, and qualities that, by 1660, were setting him and his fellow colonists apart from their contemporaries in Britain. And, after a span of only three more decades, they were being identified as belonging to a clearly recognizable variety of the original English stock. Henceforth they would be called Yankees.[1]

The reciprocal influences operating between man and nature that transformed the Englishmen of 1640 into the Yankees of 1690 were extraordinarily complex, and they evoked varying responses. The people, including their leaders, were perennially placed in novel situations and had to encounter unforeseen conditions to

* This essay is reprinted from the *Proceedings of the Massachusetts Historical Society*, Volume LXXXVIII (1976), 3–18, with the permission of the Director.

which they reacted with displays or gestures of acceptance, acqui-
escence, modification, transformation, invention, pride, humility—and also at times with fear. But whatever attitudes a man
evinced, he and his fellow settlers changed as they adapted themselves to living in New England. Sometimes the change was so
gradual as to go unnoticed; at other times it was rapid and brought
about open violence. Becoming a Yankee was an unanticipated,
unplanned adventure, and, by the very nature of things, the kind
and degree of transformation varied with individuals, were often
disturbingly uneven, and occasionally frighteningly strange. The
initial experience of the struggle to survive and the haunting sensa-
tion of uncertainty felt by most people gave way after a decade or
so to a feeling of permanence and hope, and so it continued.

Although the acknowledged goal of the first colonists had been
to attain a state of godliness, the Puritans lived and died in the
midst of life, confident that this was God's will. In every crisis,
personal or communal, they were sustained by their faith that the
Lord was on their side and that whatever happened was by His
volition, and they succeeded by means of resourceful improvisa-
tions in making an unprecedented accommodation to their harsh
surroundings. The frequent juxtaposing of a mounting individ-
ualism and the binding requirements of a "well ordered society"
can be either a good thing or a bad thing, but in the end, the New
Englanders appear to have silently agreed that what was best for
the town, the church, and the commonweal was best for the individual as well. Comments such as these may seem obvious, possibly trite. Nevertheless, the history of life in early New England
has not yet been examined in detail *from all sides* as a process of
adaptation.

The adjusting and accommodating started as soon as the emigrants embarked on the Atlantic crossing. The vast expanse of the
ocean forced a radical revision of insular ideas of distance, and long
after their arrival on the western shores, the sea continued to be
thought of as a barrier between them and their homeland.

The customary approach for the ships carrying the members of
the Great Migration to New England was from the northeast along
a great circle route. This meant that the first glimpse the passengers had of the new land was usually one of the large wooded
islands in the Gulf of Maine—Matinicus, Monhegan, Damaris-

cove—which for decades had all been visited by fishermen from the west of England. Sometimes a ship made its landfall farther to the southwest off the Isles of Shoals and then bore inshore past Cape Ann and on into Massachusetts Bay. Few of these new-comers ever failed to be impressed by the sight of the dozen or so "pretty Islands" they sailed past as their ship entered Boston Harbor.[2]

When the vessels bearing the Puritan colonists stood in closer to the mainland, the now notorious irregularity of the coast be-came evident, as well as the dense forest, which appeared to blan-ket the entire land. Not until they went ashore at Salem, Boston, or Plymouth did the new arrivals discover the many extensive open places cleared of trees and underbrush by such natural agents as wind, ice, snow, storms, or by fires ignited by lightning or inter-nal combustion or deliberately set by the natives. Divine Provi-dence, so the Puritans devoutly believed, had created these forest openings to furnish land for immediate occupation and cultivation by a favored people.[3]

The Great Migration of the English to continental Europe, the Caribbean, and the mainland of North America attained a total of about 80,000 souls by 1642. Of this number, contemporaries calcu-lated that 21,200 or somewhat more than a quarter of all emi-grants, had crossed to New England between 1620 and 1642. Be-cause of the natural increase alone, and despite the return of some people to England and serious losses in the Indian wars, the popu-lation of the New England colonies rose from about 24,000 in 1643 to 45,000 to 50,000 by 1660 and 91,900 in 1690. After 1636 or thereabouts, the standard of living started to improve and mounted steadily throughout the seventeenth century. The rigors of the climate notwithstanding, New England proved to be a much healthier country than Old England. Superior diet, good housing, ample fuel for warmth in winter, and the absence of decimating epidemics during most of the period made for widespread good health. For the same reasons, maternal and infant mortality were much lower than in the mother country, and the birth rate ran well ahead of the death rate. This, in turn, led to a doubling of the population every two or two-and-a-half decades.[4]

During the first years ashore, the inexperienced rank and file of Englishmen had to adjust themselves, their ideas, and their habits to new kinds of housing, husbandry, industrial and commercial pursuits merely to survive in the new land. At the same time they were forced to reconcile themselves to sudden social changes, to novel religious and political views and institutions, to diverging cultural attitudes, to shifting manners and unfamiliar moral standards, and to new psychological stresses and unprecedented superstitions. Moreover, they had to accustom themselves to treating strange ailments and practicing odd ways of maintaining health, as well as having to eat curious foods—even socially undesirable and radical persons, like the ordinary conforming folk, clung to the fondly remembered diet of Old England, as is true of emigrants of any epoch.

Concurrently and constantly, the Puritans had to adapt themselves to strange climatic and physical conditions. Not one of the authors of the promotion tracts published before 1640 dealt candidly with the severe climate of New England; and the first settlers studiously refrained from writing home about it. Occasionally, when three-day storms swept in off the sea, the colonists were beset by serious disasters: very few of the earliest shelters and primitive houses could long withstand a steady downpour of rain; and at times nearly every kind of structure blew down. From July to October there was always the possibility that a hurricane might blow in from the Caribbean; and not infrequently high tides took out fragile wharves and piers and drove vessels of all tonnages on shore.

Rapid changes in the weather and extremes of heat and cold not only surprised the settlers, who had been used to the milder and less variable weather of Britain, but forced them into making many modifications in their mode of living. During the summer, the prevailing winds came overland out of the southwest and often brought high temperatures and oppressive humidity. In the winter, cold west and northwest winds brought in extremely frigid weather for long periods, for winters were much more severe in the seventeenth century than they are today. The records of the United States Weather Bureau indicate that for the last 135 years the average temperature rose at the rate of three degrees a century, which suggests that the first New Englanders experienced winters averaging nine degrees colder than those of the twentieth century. Closely related is the fact that since 1898 the average annual snowfall has decreased about fourteen inches. The entire Atlantic coast from Maine to Manhattan received an almost even monthly rainfall with an annual range of from forty to eighty inches a year and experienced spring floods or freshets that took out bridges and caused extensive erosion of the soil. These extremes and vagaries of nature necessitated both physical and mental adjustments by the new colonists.[5]

Shortly after their arrival, some venturesome men rounded Cape Cod in tiny pinnaces or shallops to discover for themselves insular waters of southern New England wherein lay Nantucket, Martha's Vineyard, and the Elizabeth Islands, which Bartholomew Gosnold had visited as far back as 1602. They found that the great bay called "the faire Narragansitt" encompassed a number of "fine Islands" in addition to Aquidneck, or Rhode Island. Block Island lay about fifteen miles south of Point Judith, and approximately the same distance beyond it stretched Long Island for 150 miles. The latter forked at its eastern end into two long prongs that formed Peconic Bay, and within them lay two protected, fertile islands, Shelter and Gardiners, both of importance in early New England history. Off the mouth of the Pequot (now the Thames), Fishers and Plum islands served as stepping stones to Orient Point, at the end of the northern prong; at the tip of the southern prong was Montauk Point, the easternmost end of Long Island. To an observer on the deck of a small vessel plowing along the five hundred miles of coast, it would have been abundantly evident that most of the long stretches of the New England shores remained to be cleared and settled.

For nearly a decade after 1630, the ocean, the islands, the jagged shoreline, and the omnipresent woods were all that most of the settlers knew about the country as they busied themselves with building shelters and planting crops. That the islands, all of them situated close to New England's shores, provided many safe anchorages and places to erect flakes and stages for drying and salting the cod and mackerel caught in the Gulf of Maine or on Georges Bank before they were shipped to faraway markets is one of the better popularized activities of the region's early history. Almost unrecorded, however, especially of those islands south of Cape Cod, is their equally great value as grazing lands, protected as they were by the sea from predatory animals. North of the

Cape, certain peninsulas, such as Nahant and Marblehead, provided much the same kind of security for hogs, cattle, horses, and sheep.

Attracted by what they saw during their cruising, some of these explorers and their enterprising friends began to settle in certain of the outlying areas. From Pemaquid in Maine, they spread southward and westward along the coast to Westchester Creek in New Netherland, and on Long Island all the way from Montauk to Jamaica and Gravesend. Save at Brookfield in Massachusetts (1667), all settlements were carefully situated on rivers or small streams that afforded access to deeper waters by means of canoes and other shallow-draft craft. Rising seaports and farming communities fringed the seashore or banks of the principal waterways; and the transplanted English quickly grew adept in practicing aboriginal, as well as European, agriculture and carrying on the routine of shop and countinghouse.

Just as the first colonists had divested themselves of insular notions of distance on the long voyage to New England, Edward Winslow, John Oldham, and other daring spirits dispensed with old concepts of space when, as though driven by an onshore gale, they pushed inland. In the course of sailing up and down the coast in their small boats locating a "store of superb harbors," they also entered the lower reaches of several of the larger rivers, particularly the Kennebec, Piscataqua, Merrimack, Taunton (from Narragansett Bay), Pequot, and Connecticut. From such explorations, not only did they learn much about the topography of the seaboard lowlands, but they also became aware of rich alluvial meadows.

By 1660 a goodly number of the New Englanders, especially those of the second generation who were not circumscribed by memories of the narrow limits of Old England, had acquired by hearsay if not by actual sight and experience some understanding of the vast extent of the country. Only in the Piscataqua region, in Essex County, and in the Concord, Lancaster, and Brookfield areas of Massachusetts, in the western parts of Plymouth Colony, around Narragansett Bay, and up the Connecticut Valley as far as Hadley had the planters penetrated inland in search of fertile meadows and intervales. In 1680, the Reverend William Hubbard, while conceding that the country was generally "more mountainous and hilly then other wise, and in many places very rocky and full of stones," added, "Butt here and there are many rich and fruitfull spots of land, such as they call intervall [intervale] land, in levell and champaign ground, without trees or stones, neere the banks of great rivers, that often times are over flown by the channells of watter that run beside them. . . ." Elsewhere, most of the vast interior of New England still remained covered by virgin forest, whose denizens, except for the vanishing beaver, roamed undisturbed.[6]

From the very beginning, and increasingly as settlements spread along the coast, the offshore waters and the great rivers proved to be economic assets, for they teemed with many varieties of edible fish fit for table use or export to the Iberian Peninsula, the Wine Islands, and the West Indies. In addition, the arrival every spring of shoals of alewives to spawn in the streams that joined with the sea provided the best source of fertilizer, a practice learned from the Indians. The Atlantic Ocean, once thought of as a formidable barrier, was now a most important maritime advantage as colonial and foreign ships plied back and forth on the sea lanes, the links to Old England. Regular and essential exchanges of every imaginable sort made possible the creation of a New England three thousand miles distant from the homeland, and some Yankees, like the Englishmen from whom they sprang, became notable handlers of ships.

The emigrants from Britain had not been forewarned of the need to make adjustments in housing and to climatic conditions, but all of them knew from listening to sermons and reading tracts about New England that they would have to deal with the Indians. For more than a decade before the Great Migration, both lay and clerical Puritan leaders had been formulating their ideas about colonizing, including ways of ensuring fair, reasonable, legitimate, and humane treatment of the aborigines. Indeed the royal charter granted on March 4, 1628/29, to the Governor and Company of Massachusetts Bay specifically charged its freemen to conduct themselves so peaceably that "their good life and orderlie conversacion maie wynn and incite the Natives of [the] Country to the knowledg and obedience of the onlie true God and Savior of Mankinde, and the Christian fayth, which in . . . the Adventurers free profession is the principall ende of this Plantacion."[7]

It was most fortunate—a Divine Providence, as the Puritans

believed without question—that they landed and settled in a
region where, in 1616/17, a "plague," possibly rendered more vir-
ulent by a severe drought, had reduced the native population of
New England from approximately 25,000 in 1600 to somewhere
between 15,000 and 18,000 by the time the Pilgrims arrived in
1620.[8]

The English located their first communities in the very vicinity
where the epidemic had been most fatal—from the south banks of
the Piscataqua to the eastern shore of Narragansett Bay. There the
remnants of once powerful tribes welcomed them as potential
allies against the fierce Tarrantines (Abnaki) living north of the
Merrimack and the warlike Pequots ranging the territory between
the Pawcatuck and the Connecticut rivers. Only very slowly did
the English awaken to the profound significance of intertribal ri-
valries.[9]

Notwithstanding the care with which their leaders had prepared for the encounter with the Indians of New England, meeting
the red men face to face proved a disturbing experience for many
of the Puritans. The knowledge of the terrible massacre of the English in Virginia in 1622 caused those who crossed to America after
1629 to be ever watchful and cautious and to refrain from any
aggressive action. They entertained no racial aversion to the Indians and, their detractors to the contrary, the Puritans' missionary
motive was sincerely held by the settlers, who endeavored to implement it according to their lights. They insisted that the land
used for their settlements be paid for, ordinarily in advance, and
they quickly learned much of the forest lore and agricultural ways
of the Indians. Even so, in the early years, merely to distinguish
the members of one tribe from another turned out to be very difficult for the white men. All of the natives of New England spoke
the Algonkian language, but in so many dialects that communication between the red men and the white was severely
limited.[10]

The English soon discovered that they were dealing with "savages" and that it would be necessary to modify their plans of converting them individually because of the Protestant emphasis upon
reading the Holy Bible and the unparalleled high intellectual
requirements of Congregational Puritanism. They realized that before conversion, the natives would have to have a minimum of ed-

ucation; and to provide this and bring even a single Indian out of the "savage" state and teach him to read and comprehend Holy Writ would take much time and preparation. Herein the Puritans displayed a marked contrast to the speed with which the Spanish and French missionaries converted masses of illiterate red men to the Church of Rome. And if, as the Marian exiles had insisted so soberly—and humorlessly—"God is English," then the aborigines must first be Anglicized by learning to wear clothes and, in a dozen other ways, to lead a civilized life.[11]

Consequently, while not abandoning their hopes of converting the red men, the colonists endeavored to devise new ways and means for living amicably side by side with their Algonkian neighbors. They succeeded for the most part and, after the Pequot War of 1637, despite some friction, no major breach of the peace between the red men and the white men occurred before 1672.

The imperative need to adjust themselves to the natural environment and the importance of cultivating friendly relations with the Indians became self-evident soon after the arrival of the Puritans in New England, and the accommodations they made were generally acceptable to everybody in the settlements. Modifications of social and religious attitudes came about more slowly, and many of the settlers were barely conscious of the changes. To understand how this people, born and bred in Britain, came to abandon its Utopian dream, it is necessary to review the background and character of the band of emigrants who crossed to New England and to examine the impact of the new environment, as well as the effect of the new ideas and views they acquired later on from their fellow Puritans at home.[12]

The quality of those early Puritan emigrants who made the Atlantic crossing at considerable sacrifice was unsurpassed in the history of colonizing. With their ministers and lay leaders, the ordinary folk shared a sense of opportunity, a capacity for sound reasoning, and a conviction of divine mission; together they were going to set up a godly society. *Nil desperandum, Christo duce* could have been their motto, for, alone of the outgoing Englishmen, this large body of Puritans believed that it was carrying out God's injunction. "The universal and all-pervading influence of the Puritan vision of life as an organic and integrated whole, of which religion was at once the motive power, and the ultimate aim . . .

cannot be stressed too often or too heavily," an able Roman Catholic scholar reminds us. There was much substance in Master William Stoughton's oft-quoted insistence in an election sermon of 1668 that "God sifted an whole nation that he might bring choice Grain over into this Wilderness."[13]

This migration of superior common people deserved and got a high order of leadership from dedicated gentlemen who crossed with them in search of God, not gold nor even glory. As one of them, Sir Henry Vane, once asserted, they had "public self-denying spirits." Composed principally of members born into the lesser gentry and ministers who had become gentlemen by virtue of university training, the Puritan leadership joined religious virtues with the aristocratic creed of the English upper class. Possessed of moderate wealth to invest in colonizing, these men, to be sure, desired power, but they exercised such authority under the restraint of fixed ideas and principles. Individually and collectively they showed themselves to be skillful planners and generally competent administrators with remarkable talents for organization, which they exhibited at the very outset by forming congregational churches and new provincial governments, together with town and county communities. Recognizing the need for men of their capacities and wisdom, the middle and lower orders among the Puritans not only acknowledged their superiority but welcomed their leadership and kept them in places of trust for long periods.[14]

The uniqueness and vigor of New England society during the seventeenth century stemmed from the fact that it was the accomplishment of *a group migration*. In many instances ministers were accompanied into exile by substantial numbers of their English parishioners; more often, bands of Puritans went out together from a single town or from an area, but even more important than Puritanism was the family makeup of the population. It was the outstanding feature of life in the five colonies of emerging Yankees. The Great Migration had been a family enterprise in which husbands and wives, with their children and household servants, all ventured westward voluntarily and freely. They did not go out as laborers under individual contracts or by enticements, force, or misrepresentation, as did most of the youthful emigrants who left for the Chesapeake or the Caribbean and who lacked the steadying influence of the family.[15]

No other English colonial undertaking began so auspiciously as that of the founders of a new Canaan, who made up, as Thomas Hutchinson deduced from documents now lost, "about 4,000 families." A large majority of the people belonged to families of Puritan yeomen or husbandmen in Old England. Ranging in age from newborn babes to "ancient" men and women, more than half of them were middle-aged and experienced in life; the ratio of men to women was conspicuously closer than in the southern or island colonies. Only a few bonded youths who did not belong to some household joined the migration, and no paupers crossed with the Puritans.[16]

The enterprise, as a matter of record, was well financed—most of the emigrants had paid for the passage and had taken along some supplies of their own. Those who were unable to do so received assistance from the rich leaders of the Massachusetts Bay Company. The very sound way in which it had been organized eased many of the difficulties of adapting to a new life.[17]

In the strange, new, and rock-strewn land, idleness became for the settlers a far greater sin than even John Calvin and William Perkins had taught in the doctrine of the calling. To be idle in New England was to be antisocial, as well as ungodly. Such an offense to the community ultimately gave rise to one of the most American of proverbs: "Root hog, or die." In all the predicaments in which they found themselves, the Puritan colonists usually displayed the best English virtues, and their very Englishness enabled them to survive and succeed in building a sound society. At the same time, their ready acceptance of being a chosen folk always induced a high opinion of themselves; so high in fact that, as a group, they long remained guilty of the cardinal sin of pride; and more than once, unfortunately, the arrogance and relentless vindictiveness so characteristic of the English nation everywhere also appeared in New England.

Driven onward by practical needs, sustained by the knowledge that God was with them, very few of the rank and file of the New Englanders had ever entertained John Winthrop's notion about erecting "a Citty upon a Hill." Heaven on earth was for them merely a dream of the future; freed as they were from many of the vexations and troubles that beset kinsmen and friends in the homeland, they worked long and hard to create out of the forest a decent, moral, hospitable society of like-minded Puritans. The idea of progress as the key to the philosophy of history had not been adumbrated in 1660, but something like it, *improvement of society*, if you will, seemed patent to those persons who, like Edward Johnson in 1650, looked around them in New England and with amazement compared it with the wilderness they recalled of 1630.[18]

In England before 1629, the Puritan leaders had begun the process of adapting their followers to sudden religious, political, and economic changes. Their view of life was inspired, idealistic, and otherworldly at times despite their inbred practical bent, and in New England the common-sense, down-to-earth, often stubborn mood of the "silent Democracy" not infrequently interposed a healthy realism to the Utopianism of this "speaking Aristocracy," which wisely, and on the whole generously and unselfishly, ultimately accepted such adjustments. Not even John Winthrop or John Cotton ever approached success in building a Utopia. Because all of them had in good measure what Edward Johnson cogently labeled "Christian Courage," the Puritans and their leaders manifested a resilience that enabled them to face all obstacles and to surmount most of them.[19]

Mass immigration suddenly ceased in 1642, and for a time more Puritans sailed homeward "in expectation of a new world" than came over to the new land. Such additions as the population received thereafter, other than by natural increase, were of individual Englishmen, not groups—save for a few hundred Scottish prisoners shipped over by Oliver Cromwell in 1652.[20]

Aside from a severe temporary economic shock, the ending of immigration in 1642 proved conducive to the growth and stability of the colonies. During the next twenty years, the inhabitants of all of them, from the Kennebec River to Westchester Creek, predominantly Puritans from nearly every county in England, enjoyed an uninterrupted period in which to root themselves, grow accustomed to each other's speech and ways, and raise a generation of children native to the land without any interference from the outside.

The only nearly complete transfer of English society to the New World during the seventeenth century was to be found in

New England. Its inhabitants were more purely English than those of the Chesapeake or Caribbean colonies of England; they felt no need to mongrelize. In New England the speech of the Thames Valley—the King's English—gradually began to prevail over the dialects imported from Cornwall, Devon, and Yorkshire. With the exception of Maine and the basin of the Piscataqua, where Anglicans and Nothingarians were probably in the majority, uniformity in religion also nourished social and political unity; even the maligned Rhode Islanders were puritans and, as the years passed, their differences with the Congregationalists of the New England Confederation proved neither fundamental nor lasting.

An understandable emphasis upon the geographical isolation of the Puritan colonies has led many writers to overlook certain profoundly important links between New England and the mother country. During the first decade, the steady arrival of shiploads of immigrants kept the older settlers fairly well abreast of developments in Britain. From 1642 until 1660, they did experience an isolation of sorts, but even so, as Old England changed under the pressure of events, many men sailed eastward to fight in the Good Old Cause, and it seems evident that, through them, their friends in Parliament, and the many industrious writers of long letters among the Independent ministers, the men of New England learned more about what was going on in Britain and on the Continent than did their Royalist brethren of the Chesapeake and Caribbean dominions from the entourage of King Charles I.

Although the ocean highways from Boston to the British Isles, to Europe, and to the West Indies were thousands of miles long, they served as avenues of transport, not merely of goods but of ideas and news that brought the New Englanders (sometimes reluctantly) into the Atlantic Civilization. Commerce was a most civilizing and broadening activity in this age, and it impelled the colonists to discard some of their provincialism and even to soften or modify some of their harshest religious attitudes. Overseas, almost without becoming aware of it, the merchants from New England encountered and absorbed some of the new and revolutionary ideas then circulating, and sooner or later their clerical and political associates had to adjust themselves to these ideas. In the process, the "New England Way" was ineluctably altered.

Puritanism did not fail or degenerate in New England, as so many critics have maintained; indeed, as William Hubbard declared at the time, it was "an incredible successe," which has never been matched. After 1660 the emerging Yankees assumed control of affairs and gradually leavened the imported strict Millenarian faith of the first leaders of the Puritans. Thirty years later, old Joshua Scottow might lament the cooling off of "the primitive zeal, piety, and holy heat found in the hearts of our parents," but the Yankees of 1690 retained what remained—and it was no small amount—and added a saving ingredient of native humor for transmission to future generations.

English Puritanism itself had had to be accommodated to the new societies in the New World in the very beginning and modified frequently thereafter. For New England, this was "a good thing," because "a Citty upon a Hill" had never been desirable for, nor wanted by, the great body of the colonists of Massachusetts, Plymouth, Connecticut, and New Haven, let alone those in the three other settlements.

Knowingly or unconsciously the New Englanders made major and minor concessions in all areas of life. Far from being rigid, they appear amazingly flexible, supple, even compliant. Failures were relatively few, and they managed to survive not just the extremes and caprices of nature that had necessitated both physical and mental adjustments but to develop a decent if frugal mode of living enjoyed by few English colonists in other climes. The essential Yankee mood was to trust in God and make a virtue out of necessity by means of unceasing toil, and by 1690 these Yankees had accomplished an almost complete adaptation. Such changes may have seemed dictated by Divine Decree at the time, but they profoundly and irrevocably transformed the characters of the right New England men and women and made them into Yankees.

And I repeat here what I have written elsewhere, that the history of the last five centuries does not record—if indeed it ever had previously—"a colonial enterprise as successful as the one that certain of the English Puritans [their many limitations and faults notwithstanding] undertook of their own volition in seventeenth-century New England." But let us end with what the leaders of New England, gathered at the synod of 1679, had to say about the achievement of right New England men over a half-century: "If we look abroad over the face of the whole earth, where shall we see a place or people brought to such perfection and considerableness, in so short a time? Our adversaries themselves being judges, it hath not been so with any of the outgoings of the Nations."[21]

IV

Yankee Use and Abuse
of the Forest
in the Building of New England
1620–1660 *

I

A safe refuge in the New World where Puritans could escape the domination of the hierarchy of the Church of England was the primary attraction for the founders of the Massachusetts Bay Company who planned to settle in New England. As the news of the proposed colony spread, many ordinary people, who had no land they could call their own and who suffered from the lack of adequate housing, fuel, and a balanced diet, joined in what became known as the Great Migration.

"There is nothing to bee expected in New-England but competency to live on at the best, and that must bee purchased with hard labour," Master John White of Dorchester candidly announced in *The Planters Plea* (1630). In the colony he and his associates proposed to establish, almost the only source of wealth would be found in or on the land. For many of them, as well as for the majority of their fellow Puritans, even a tiny plot of ground was both too costly and too hard to acquire in England, but in America, land was plentiful.[1]

On March 3, 1628/29, the day before they received the charter

* This essay is reprinted from the *Proceedings of the Massachusetts Historical Society*, Volume LXXXIX (1977), 3–45, with the permission of the Director.

from King Charles, the officers of the Company began to discuss "how some good course might be setteled for the devision of the lands" of their grant "whereby an equallety may bee held, to avoyd all contention twixt the adventurers." These prescient Puritan gentlemen knew about the evils that had resulted from the gathering of the "means of wealth" into the hands of a very few men in Virginia and the British islands of the Caribbean. They had no intention of selling off or granting large tracts to enrich a few favored individuals infected with the prevailing land-hunger of the English nation.[2]

Assured of a seemingly inexhaustible supply of land, which was guaranteed in their patent from the Council for New England, the Court of Assistants approved a land policy on May 21, 1629, that was unique in the annals of colonization. No land was to be sold; rather it was to be given away in comparatively small parcels to the many men who ventured their all in migrating: even the humblest landless freeman from England would receive at least a fifty-acre tract in the new society. To masters of families who went out on their own, the Company offered fifty acres for themselves and fifty acres for each of the relatives who accompanied them, together with additional land "according to their charge and qualitie" if the Governor and Council in New England should deem it appropriate. Furthermore a master would be allotted fifty acres for every servant he transported. These stipulations were consonant with the Puritan view of equality of landholding: that it should vary with the station, needs, and capacities of individual colonists.[3]

In framing the policy of granting land to right-thinking people, the leaders of the Puritan Hegira exhibited a rare and penetrating understanding of both the religious and material aspirations of Englishmen. "Intentions are secret, who can discover them," John White freely conceded: "Necessitie may presse some; Noveltie draw on others; hopes of gaine in time to come may prevail with a third sort; but that the most, and most sincere and godly part have the advancement of the *Gospel* for their maine scope I am confident." His remark that "nothing sorts better with Piety than Competency" is the clue to why the planters thought to present an irresistible opportunity to the thousands of troubled and landless men by coupling the offer of free worship with free land: the promise that in New England one could not only enjoy freedom of worship but also hold broad acres in fee simple on which to build a house of his own.[4]

II

It is next to impossible for anyone in the twentieth century to have any comprehension of the nature and extent of the problems facing the immigrants who landed during the first decade of the founding of New England. When the "multitude of people" from the twelve ships of the Winthrop Fleet, estimated at about fifteen hundred, went ashore with their effects at Charlestown in June and July 1630, their first concern was to arrange for the erecting of about three hundred temporary shelters. Neither they nor the remainder of the 21,200 souls who arrived in Massachusetts before 1642 could expect to enjoy "Christ and his Ordinances in their Primitive Purity" until they ensured their physical survival in "a desart Wildernesse."[5]

Shelters were, understandably, hastily put up and primitive, whether for immigrants landing for the first time in the New World or, later on, for those "going out" from the settled areas around Boston, Salem, and Plymouth to establish new plantations in the wilderness woodlands. Not a few of the poorer immigrants of 1630 lived in tents well into the autumn; others, in great haste, imitated the Indians and threw together wigwams—some of them might have been prompted by Christopher Levett's account of the winter of 1623–24 which he spent by the shores of Casco Bay: "We built us our wigwam, or house, in one hour's space. It had no frame, but was without form or fashion, only a few poles set together, and covered with our boat's sails, which kept forth but a lettle wind, and less rain and snow." Huts and wigwams sheathed with bark and lined with mats or skins procured from the natives provided slightly more warmth, but most families, having no experience or skill in such work, often spent months in cruder and less watertight booths and shelters.[6]

A much improved temporary dwelling, according to Edward Johnson's epitome of a method used at Concord, was improvised by 1636: when colonists find a "place of aboad," they "burrow themselves in the Earth for their first shelter under some Hillside, casting the Earth aloft upon Timber [to form a roof]; they make a smoaky fire against the Earth at the highest side, and thus . . . provide for themselves, their Wives, and little ones, keeping off the short showers from their Lodgings, but the long ones penetrate through to their great disturbance." At the age of six, Michael Wigglesworth came over with his family in 1636 to Charlestown; after a stay of about seven weeks, they sailed in October south to New Haven. Years afterward the poet recalled vividly how "we dwelt in a Cellar Partly under ground covered with Earth the first winter. . . . One great rain brake in upon us and drencht me so in my bed being asleep that I fell sick upon it." These cellars were seldom found in English cottages and may be counted as one more adjustment or adaptation to New England conditions. Besides being the warmest of all temporary living quarters of that time, they also served as foundations for later permanent structures. They were not uncommon in Connecticut and at New Haven as late as 1646 and may be considered as predecessors of the sod houses on the western prairies so vividly described by Rölvaag in *Giants in the Earth*.[7]

In conjunction with the policy of giving away land free of all encumbrances, planners in England directed the Governor and Council at Salem to allot tracts to the settlers on the same acreage where they expected "to inclose and manure, and feed their cattle." To achieve this, the land was to be conveyed to "each particular man" in the name of the Company. As a result, the families of every new settlement in New England in the seventeenth century, as soon as conditions permitted, could go about raising tiny cottages and move out of their primitive shelters.[8]

The never-ending need for more houses caused building to rank second only to farming among the economic activities of early New England; indeed during the first twelve years it must have absorbed more of the time and energies of the colonists than the planting and growing of Indian corn, rye, and oats. More than seventy ships, each carrying an average of 100 passengers, arrived during 1630–31, and in 1635 alone about 3,000 people came over; even in 1640 about 1,650 landed. Consequently up until 1642, when the Puritan immigration ceased, some kind of housing was required to accommodate approximately 24,000 souls that meant the erection of between 4,800 and 5,000 dwellings of all kinds.

After 1643 the inhabitants began to catch up with the demand, and thereafter, at least until 1690, other requirements kept construction in third place—just behind the lumber industry but well ahead of shipbuilding.[9]

The feverish pace of building was observable in all the Puritan colonies. By 1635 possibly two thousand small houses had gone up in the plantations around Massachusetts Bay, nearly all of them patterned after the ones erected in Plymouth Colony prior to 1624 and at Salem after 1629. Salem contained 226 dwellings in 1638 and about four hundred by 1660. The first months of 1636 were spent in "accommodating" the three thousand "new come Guests" of the previous year. In June 1637, just two years after the colony and town of New Haven were founded, David Pietersen De Vries, who was sailing up Long Island Sound, arrived there and discovered that the English had begun to build a town on the mainland and already about three hundred houses and a fine church had been erected. On his way back to New Amsterdam, the Dutchman saw more than fifty dwellings going up on the south bank of the Housatonic at the site of the present Stratford and comparable activity at what are now Norwalk, Stamford, and Greenwich.[10]

In most of the new plantations, entire populations needed temporary shelters; the town of Dedham was an exception, for many of its inhabitants had previously lived near by. By going over to the site almost daily between the spring of 1637 and the summer of 1638, they managed to fell trees, hew beams, and to frame and clapboard their new dwellings before they actually moved into them. The vast majority of houses built in the outlying parts were tiny one-story cottages with thatched roofs.[11]

These cottages had frames of square-hewn timbers to which a sheathing of overlapping clapboards or one of boards, laid flush, was attached. To provide insulation, the space between an interior wainscoting and the outer sheathing was filled with clay and chopped straw, sometimes made into rolls called cats. In the lower part of his garden, Robert Keayne, a substantial merchant and selectman of Boston, still kept the "Mud Wall house" he had occupied soon after his arrival in 1630. In such a structure the corner posts were connected with crosspieces between which wattle-and-daub was filled in, as in East Anglia and Essex. In the absence of lime or other suitable binding material, this celebrated half-timber construction would not withstand driving northeast storms and soon fell into disuse.[12]

The ground floors of most cottages ordinarily consisted of only one room with an interior porch (entryway) in one corner from which a steep staircase or a ladder gave access to a "loft" formed by the gable roof above. The hall, as the English called the one all-purpose room, had as its principal feature a huge fireplace. Both the fireplace and the massive chimney that rose out of it had heavy timber frames filled in with cats heavily daubed with clay. Finally, a roof of thatch made from the tall straw, which grew wild near by, covered the cottage.[13]

Men with weak constitutions ought to remain in Britain, William Wood sagely advised in 1634, "for all new England must be workers in some kinde," as many Englishmen learned when they reached the New World and started to build houses. Colonial husbandmen seldom possessed any folk skills with the felling ax, broad ax, adze, and frow; and only a few of them had ox teams to haul the heavy beams from the spot where they had been felled. Riving clapboards or shingles from bolts of wood, even when they acquired the necessary tools, was a genuine novelty; and thatching was a trade that demanded thoroughly trained workmen. In fact, about all the untutored colonist could contribute to the building of his house was the muscle to help in such ordinary tasks as carrying timbers and boards and assisting a carpenter to raise the frame. Only slowly and clumsily over the century did the Puritan develop into the Yankee jack-of-all-trades.[14]

Though fully aware of their shortcomings, the colonists generally assumed that the houses they struggled so hard to build would endure for years to come, but all too soon they found them to be poorly constructed and far from weathertight. The hewn timbers and boards used in the early years were unseasoned—time did not permit such a refinement—and green wood, fresh from newly felled trees, the sawpit, or clapboard bolt, soon warped and twisted. Not infrequently house frames were made of flimsy stuff by these amateurs, and poorly thatched roofs inevitably leaked. Hugh Peter of Salem mentioned casually in 1638 that "Mrs. Beggerly's or rather Mr. Skeltons house, which is now falling to the ground, if something bee not done" was but ten years old. The scarcity of any workable building stone and the lack of any lime other than that made from oyster and clam shells dictated that the nogging used in timber-framed walls and chimneys be of the catted or daubed variety. When great winds, such as occurred in the hurricane of 1634, swept over New England, many of these unsubstantial structures collapsed; the following year a northeast storm "passed through the whole country overturning sundry houses, uncovering divers others." Lucy Downing wrote her nephew John Winthrop, Jr., late in 1648 that the Reverend Nathaniel Norcross and Edward Godfrey were urging her family to move to Agamenticus (York); Mr. Godfrey promised them "a hows with 3 chimnyes . . . if 2 of them blowe not down this winter, wich may be feard, being but the parsons howes."[15]

The need for the services of trained housewrights—men able to hew and square beams, cut mortices and tenons, and then assemble the frames—was soon apparent. Although some came among the crowds now arriving from England, there were never enough to meet the demand. They were retained to build houses for the ministers, or else rich settlers monopolized them. Furthermore, their wages rose to such a point that the ordinary husbandman could not afford to hire them.[16]

The price charged by housewrights for their services and the desperate need of the settlers for expert assistance in putting up their houses led the Court of Assistants as early as August 1630 to regulate wages. Carpenters, joiners, sawyers, masons, bricklayers, and thatchers were ordered not to charge more than two shillings a day, and no one was to pay more than that rate. In March of the next year, the Assistants, recognizing their inability to enforce the orders, freed wages. Governor John Winthrop observed in October 1633 that workmen had raised their wages "to an excessive rate"—carpenters were asking three shillings and six pence. In that same month the General Court, influenced by complaints that many workmen loafed on the job and that others refused to work a full week, ordered wages reduced to fourteen pence a day. These orders proved no more enforceable than the earlier ones, and three years later the General Court turned the regulating of wages for craftsmen, laborers, and servants over to the towns, whose selectmen were expected to set wage schedules as the courts of quarter sessions had done in England. This measure seemed to improve the situation for a time, "but it held not long."[17]

Massachusetts was not the only colony to be plagued by high wages and high prices, for the General Court of Connecticut, judging in 1641 that the artificers had become "a law unto themselves," forbade carpenters, ploughwrights, wheelwrights, joiners, coopers, masons, and smiths to charge more than twenty pence a day from March to October or more than eighteen pence for the rest of the year. They also ordered them to work a fair eleven hours a day in the summer and nine in the winter, not including time spent in sleeping and eating. Throughout the years up to 1660, one heard universal complaints about the "treble charge of building in this populated desart, in regard of all kinds of workmanship."[18]

III

A marked shift in the public mood, first noticed in the spring of 1636 occurred when the settlers "began to be perswaded they should be a setled people." It profoundly influenced their attitude toward most material concerns, and in the next two decades, as men made plans for more comfortable, spacious, and permanent dwellings, a veritable building boom developed. It was induced not only by increased immigration but by the need to repair or replace many of the existing, inadequate structures, to add a room or lean-to to make space for growing families, and to provide new houses for young married couples.[19]

Of prime significance in achieving the goal of better housing was the entry of more trained artisans and craftsmen into the building trades; they in turn were taking on youths as apprentices who were attracted by the prospect of high wages and living away from parental control. At a magistrate's court in Hartford in 1637, Francis Stiles covenanted to teach three of his servants the trade of carpentry four days a week during the period of their service: they were to saw and split boards and, with their own hands, supervised by Stiles or some other workman, frame them. In this same year, John Goyt went to Marblehead (where he lived in a wigwam until he could get a house) and worked for Thomas Bowen, probably as a pitman "to help saw plank." In 1639 we find the record of Richard Gridley of Boston turning over the indenture of his apprentice William Boreman to William Townsend, thatcher. Cer-

tainly trained thatchers proved to be one of the greatest boons to
people building houses.[20]

The founders of Newport had been among the richest men of
Boston and knew the labor situation there very well. Sometimes
when they were in need of workmen, they were able to persuade
apprentices in the Bay Colony to leave their masters and work for
them by promising higher wages. There were also occasions when
Massachusetts men contrived to profit by sending their servants to
work in Rhode Island, while they themselves collected the wages
and paid the servants at the Massachusetts rate. One of these men
was Richard Iles of Charlestown: he dispatched his servant, Ambrose Sutton from Oxfordshire, to Aquidneck to work with the expectation that he would pick up the wages amounting to £3 12s.
Sutton, however, had his own ideas and in 1639 absconded with
the money. All Iles could do was to ask Isaac Allerton to try to
locate Ambrose and return him to Boston.[21]

New equipment and materials, as well as trained workmen,
were necessary to produce durable, weathertight houses. In the
first years of settlement these items had been either unavailable or
too costly, but gradually they became accessible. The increase in
the supply of draft animals, mostly oxen, facilitated the hauling of
heavy timbers to building sites, for each year the forest receded
farther from the settlements. In Connecticut limited supplies of
tiles and slates were manufactured for use in roofing, and though
no limestone was discovered in Massachusetts until 1690, lime
kilns were established. All this Edward Johnson celebrated in
1648:

> Plenty of oysters overgrown the flowed lands so thick,
> That thousands of lime are turn'd, to lay fast stone and
>   brick. . . .
> Salt, sope and glasse, Tiles, lime and bricks are made,
> With orders for well-ordering of each Trade.

Each year, also, merchants imported a greater variety and larger
number of tools, locks, hinges, quantities of nails of all sizes,
together with miscellaneous hardware. The introduction of these
items and better materials made for sounder construction and further adaptations in building methods.[22]

As the demand for more houses increased, the production of bricks accelerated. Thomas Ames began to operate a brick kiln at Dedham in 1641, and the town authorized Ralph Day to use the clay pits for the same purpose in 1650. The "brick kilns in the plains" at New Haven belonged to Thomas Gregson in 1644, and a year later Edward Chipfield received permission to make bricks under West Rock. When these other communities began to produce bricks, many of the new buildings rose on solid foundations, and the use of brick and stone construction made it possible to move chimneys into the center of the houses, thereby conserving heat. John England of New Haven contracted in 1646 to underpin a new dwelling house, "make a backe to a chimney" with bricks, and to "stone" a well for £3. William Pringle, the official sweeper of the brick and stone chimneys of New Haven about 1654, agreed to serve a second year if the town would supply him with a "canvis frock and hood to cover his cloathes." Some idea of the progress made in brick making can be gathered from the report at Northampton in 1659 that Francis Hacklington burned 50,000 bricks for the house John Pynchon was building at Springfield.[23]

Walls and foundations could now be constructed so that they lasted longer, but even in the old established parts of New England, all houses, with a few exceptions, were still roofed with thatch. Upon first "setting downe," the inhabitants of the coastal towns of Massachusetts often designated as much as twenty acres of land situated between the low-water mark and the salt marshes "to bee improved for thatching houses." Properly made, a thatched roof was watertight and durable as well as seemly. On the other hand, straw ignited easily, especially in the areas close to catted chimneys, which frequently blazed out because of insufficient daubing or foulness. Isolated rural buildings were occasionally struck by lightning during electric storms—the lightning and thunder were a new and frightening experience for the English. When Constant Southworth of Duxbury was returning home from Sabbath-evening exercise on September 11, 1653, a bolt of lightning "brake and shivered one of the needles [posts] of the katted or wooden chimney" and sent splinters forty to fifty feet from his house that ignited the thatch of a lean-to. So many fires occurred at Rowley that in December 1654 the selectmen ordered that "all thatched Chimnies in the towne shall be swept" within a fortnight, and within a month the chimneys of all houses with thatch or clapboarding. Sooner or later nearly all New England communities required every householder to keep a ladder ready for use in putting out thatch and chimney fires.[24]

After 1640, when immigration subsided, the acute housing shortage was somewhat relieved; but though fewer new settlers were arriving in New England, the population increased. Many farmers and villagers prospered and, as their families grew, they needed more space. Consequently building and rebuilding never ceased, even in the most settled parts of New Hampshire, Rhode Island, and Connecticut, as well as in Plymouth and Massachusetts Bay. When Thomas Wilson died at Exeter in New Hampshire in 1643, he left his widow "my dwelling house and new frame." Seth Grant of Hartford bequeathed a house in 1646 with a "parlour" and "lodging room." And John Wilcocks of the same town left his widow "my ould howse to dwell in and that little Closett that is betweene the ould howse and the new howse."[25]

With new construction, additions to, and repairs of old houses going on apace, it was obvious to anyone interested in business affairs that there was money to be made in these years in speculative building, not only in Boston and the older communities but also in many of the new plantations as well. As early as 1637, James Luxford, steward to John Winthrop, sought guidance concerning the house he proposed to build for the governor "either to sell or keepe." The steward knew that Winthrop did "delite in playnnesse" but he hoped to contrive a plan that, at no extra cost, could also please those who "delite in commodious neatnesse . . . if for a matter of ten pound charge a man may make it happely 50 *li* better it weare cost ill saved, beside men doe now build, as lokinge on a setled Commonwealth, and therefore, wee looke at Posteryty and what may be usefull or profitable for them."[26]

Edward Johnson portrayed these early real-estate projectors in *Good News from New England* (1645), disclosing in the process that not all Puritans were humorless and incidentally that he possessed some skill in writing baroque verse. No sooner do immigrants with money come ashore than

> Five, seven, or nine old Planters do take up their
> station first

Whose property is not to share unto themselves
    the worst.
Their Cottages like Crows nests built, new com-
    mers goods attain,
    For mens accommodations sake, they truck
    their seats for gaine,
Come buy my house, here you may have much
    medow at your dore
'Twill dearer be if you stay till the land be
    planted o're. . . .
Sure much mistaken, towns have been for many
    have made prize,
Get all they can, sell often, than, and thus old
    Planters rise.
They build to sell, and sell to build, where they
    find towns are planting.[27]

Many an ordinary man found it to his advantage to buy and sell
a house if he needed more space, was changing locations, or if he
lacked the skills to make repairs on his own house or build a new
one. Such a man was Thomas Paynter of Boston, a joiner who was
going to move to Newport; he sold his house and garden in 1638 to
George Barrell, a cooper, for £28. A house, garden, and six acres
of arable land in Braintree were sold in 1639 by Thomas Brownell
to Deodat Curtis for £38; Brownell also agreed to deliver about
one thousand feet of boards. The house William Palmer of Straw-
berry Bank bought from Goodman Chatterton, cooper, he resold in
1650 to Thaddeus Riddow for £14, of which £6 were to be paid in
pieces of eight and £8 in goods; Palmer also promised to "lath and
doake the Chimney and to make it sufficiente."[28]

William Coddington, "gentleman," sold his house in Boston,
garden, orchard, pasture, outhouses, yards, ways, and easement, a
right in Spectacle Island, and a farmhouse and outhouses at Mt.
Wollaston to the merchant William Tyng in 1639 for the sum of
£1,300. This transaction may possibly have been what inspired
Edward Johnson to write as he did about "old Planters," as well as
his exaggerated appraisal of housing in New England, which he
wrote in 1651: "the Lord hath been pleased to turn all the wigwams, huts, and hovels the English dwelt in at their first coming

into orderly fair, and well-built houses, well furnished many of them, together with Orchards filled with goodly fruit trees, and gardens with variety of flowers." Sundry other observers singled out Ipswich, Cambridge, and New Haven as especially attractive villages, noting particularly the "well ordered" and "compleat" streets in the last two towns.[29]

The spread of settlements and the care taken in planning some of the new "fair, and well-built houses" confirm the unshakable confidence the Puritans had in the permanence of their settlements. The inexperienced agent and servants of Governor John Winthrop began to erect a stone building at his farm on the Mystic River near Medford in 1631, but because the stone had been "laid with clay for the want of lime," two sides of the unfinished structure were washed away during a violent northeaster that lasted twenty-four hours. Six years later, when his eldest son prepared a survey, there were five buildings at "Ten Hills," one of them a two-and-a-half-story house with a great central chimney, all of them probably of wooden construction.[30]

The following year, 1638, Samuel Symack of Ipswich wrote out in considerable detail a description of the size and kind of dwelling he wanted the carpenters of John Winthrop, Jr., to erect for him. Among other things he insisted that all doorways "be soe high that any man may goe upright under" them. He desired "shutting draw windowes" and strong "side bearers" for the second story, in which he proposed to store corn.[31]

At Boston in September 1640 John Davis, a joiner, agreed to build for William Rix a framed house 16 feet long by 14 feet wide with "a chamber floare finisht, summer beam, and ioysts" and a cellar floor. The roof and walls were to be made with clapboards, and the chimney was to be framed with hewn timbers without daubing. For his work Davis was to receive £21; all the stipulations were included in a contract drawn up by Thomas Lechford, notary.[32]

Finer materials were now sought by prosperous settlers to adorn their houses as well as to make them more durable. For more than a decade after 1630 only a very few rich immigrants could afford to import glass for windows: wooden shutters or oiled paper had to serve most people. After 1641 Richard Russell of Charlestown and several Boston merchants regularly imported glass panes packed in cribs along with small grooved lead bars called "cames" in which to set them. There were other instances when leaded glass came in sheets, which local glaziers cut to size. Most casement windows had diamond panes—the earliest surviving casement, which comes from the house William Coddington erected at Newport in 1641, has 4-by-6-inch lights. Some dwellings in Cambridge and Boston were roofed with tiles, and in 1656 the making of tiles was sufficiently developed in the Bay Colony to require official regulation. More and more, however, shingles rived from bolts won popular approval both for roofing and often for siding in place of clapboards. In the specifications for the minister's house at Norwalk in 1653, two housewrights agreed to "hange the shinckles with pinnes" on the sides of the structure and to coat them on the inside with clay for insulation.[33]

In keeping with prevailing English ideas about rank and display, the richer Puritan gentlemen, who needed more living space for their growing households and could pay for greater comforts, were able to sequester the best available local and imported building materials and command the services of skilled housewrights or carpenters to raise "mansions" for them. William Coddington recalled in 1672 that, at the settling of Boston, "I builded the first good House," which Governor Bellingham and Thomas Brattle later occupied. From the report of the sale of that structure in 1639, it was certainly a mansion in the terms of those days, and when Coddington and other Antinomian gentlemen moved south to Newport, he built a second one. It had two-and-a-half stories, a jetty or overhang on the front side, and a stone-end with a massive stone chimney. So sturdily built was this house that it remained in use until it was pulled down in 1837. Within a year of the founding of Newport, Henry Bull built a one-room stone house with an end chimney and a gambrel roof, thereby inaugurating the Rhode Island custom of using excellent local stone and lime in construction.[34]

Every New England town boasted of at least one good house in which the minister lived. The clergy, in the light of their university training and their prime importance to the community, always rated with the gentry, and they were allotted land by the Company "according to their charge and qualitie." Far up the valley of the Connecticut and miles westward from Boston, the inhabitants of the new town of Agawam (later Springfield) paid out a total of £40 between 1636 and 1638 for the building of a four-room house with double chimneys for the Reverend George Moxon. The frame measured 35 feet by 15 feet and, there was a porch 5 feet "out" and 7 feet in width "with a shady over head." Stairs led down into a cellar (a rarity in England) and up to a chamber. Goodman Burr thatched the roof for £18, while John Allen provided the thatch, nails, laths, and hauling for £3. Henry Smith charged £8 for "daubing the house and chimneys, underpinning the frame, and making the stack and oven 7 feet high."[35]

The house for the minister of the infant village of Natick was unique in that it was erected by the Indian inhabitants at the instigation and under the supervision and from a design of the Reverend John Eliot. "We must also of necessity have an house to lodge in, meet in, and lay up our provisions and clothes, which cannot be in wigwams," the Apostle of Roxbury wrote on October 21, 1650. Therefore, he reported, "I set them . . . to fell and square timbers for an house, and when it was ready, I went, and many of them with me, and on their shoulders carried all the timber together, etc. These things they chearfully do; but this also I do, I pay them carefully for all such works I set them about, which is good encouragement for labour."[36]

A year later, when Governor John Endecott and John Wilson rode out with a cavalcade of twenty horsemen to inspect this unusual enterprise, the Boston clergyman was amazed and delighted. "The Indians have built in the English manner high and large (no English hand in it save that one day or two they had an English Carpenter with them to direct about the time of rearing), with chimneys in it." Large for the day and place, the structure measured 50 feet by 25 feet and had two full stories: the first floor had a room for assemblies and another for a school; the second floor had quarters for Eliot on his biweekly visits, and a storeroom. The accomplishment of the redmen also staggered John Endecott: "To tell you of their industry and ingenuitie in building of an house after the English manner, hewing and squaring of their tymbers, the sawing of the boards, and making of a Chimney in it, making of their ground sells and wall-plates, and letting in the studds into them artificially . . . is remarkable."[37]

Undoubtedly many small one-story houses with thatched roofs continued to be built, such as the one a Massachusetts town put up in 1659 at a cost of £6 for a pauper named Alex Knight, but at the same time substantial farmhouses or rural dwellings could be seen everywhere south of the Merrimack River and west as far as the Connecticut River. At Springfield in 1659–60, John Pynchon provided himself with a residence fitting for the first lord of the Connecticut Valley. The building (42 feet by 21 feet) was joined to his old dwelling and lean-to, giving the owner possibly the largest mansion outside of Boston and Newport. The new house was sheathed by Samuel Grant with 1½ foot shingles laid on close-fitted boards; and the same carpenter also made and hung wooden gutters, a rare feature in a country house of that date. To improve the appearance of the mansion, Pynchon had Grant "Scallop" two layings of shingles on the foreside and two courses on each end of the house. For felling fifty-six trees, hewing and framing the timbers, besides hauling 156 loads of stone, Roland Thomas received £11 8s. Goodman Griswold performed the brick and stone work, including chimneys for the old house, and Goodman Wright charged £2 5s for eighteen days of sawing.[38]

New Englanders had also been establishing themselves on Long Island. Andrew Messenger and Richard Darling agreed with the town of Rustdorp (Jamaica) in 1661 to build a house for the minister, the town to provide the nails, iron hinges, clapboards, shingles, and "sawn boards" and to pay the workmen £23 "after the English account" in wheat at 6 shillings a bushel and Indian corn at 3 shillings and 6 pence. The dwelling was to be handsomely made with two chimneys "well slatted." John Budd from New Haven built the oldest structure still standing in eastern Long Island at Southold in 1649; ten years later he gave it to his daughter Anna and Benjamin Horton for a wedding present. Horton's brother Joseph, a house carpenter, took it down, moved the frame (40 feet by 20½ feet) to Cutchogue, and set it up again for £20 in pine tree shillings of Boston. It is notable for a great central chimney stack with a tall pilastered top made of locally burnt brick. Although its frame is clearly of a New England or English type, the house has an odd roof that suggests some Dutch influence, for by the mid-century, national building traditions were mingling.[39]

IV

The appearance of all permanent structures, large and small, put up in New England before 1660 can be traced to the Gothic building tradition of the late Middle Ages—it would be improper to speak of even the better mansions as architecture. Upon first coming to New England, the settlers instinctively tried to duplicate the English phase of what I have elsewhere labeled the International Vernacular, a style that can still be traced eastward from Chester across the Channel at least as far as Königsberg. For many decades a scarcity of timber had impelled builders to leave the structural framework exposed particularly in the interior of a new house; in Britain, the external half-timbering and jetty construction, known today as "black and white," prevailed all over East Anglia, Essex, and Kent and had led to the substitution of brick and daub as filling between the timbers and plaster as a coating. For some unknown reason those Puritans who emigrated from Essex and Kent and who were familiar with the older houses sheathed with weatherboards, immediately referred to them as clapboards when they settled in America.[40]

In New England the lack of skills, shortage of time, and costliness of labor rather than any Puritan aversion to beauty prevented the owners of new houses from having them embellished. With the passage of time, however, brick and stone chimneys, some of them rather elaborate, lent a decorative quality to new houses as did the introduction of the English jetty. John Norman of Manchester contracted in 1657 to build a house for the use of the minister at Bass River (Beverly). It was to be 38 feet long and 17 feet wide and 11 feet in the stud, with a porch 8 feet square, and the second story was to be "jetted over one foote each way to lap the floores below." There were to be two fireplaces on the first floor—one in the hall, one in the "stodie"—and one in the upstairs chamber. Altogether, with a shingled roof, this parsonage represented the latest in materials and construction.[41]

The permanent dwelling houses erected before 1660 typify the marked modification that the English Vernacular underwent in the forest setting of New England. The traditional European framing, fenestration, and thatched roof were retained, but clapboarding quickly and universally displaced the exposed half-timbering with mud walls; and a wooden-framed catted chimney was devised for Massachusetts and those parts of Connecticut where brick, clay, and lime, and easily worked building stone were lacking. Hitherto unnoticed by architectural historians is the fact that additional significant adjustments of the familiar mansion style to meet certain needs on the local scene—in size and shape—had to be made to produce several kinds of buildings. All of these structures may be said to constitute a variant all-wooden style, one that we may properly call the Forest Vernacular of New England.

Close by the first one-room cottage he put up for his family on his village lot, the husbandman soon began to raise a barn. This structure generally preceded any more elaborate housing. As he cleared his acres and brought them under cultivation, he needed a barn for the threshing and storage of corn and hay. Building of small barns for yeomen had been going on in England since 1590, and because most of them were of wooden construction, the settlers had a variety of examples to guide them in the new land. The long cold winters and the deep snow, however, posed the additional problems of shelter for livestock and a protracted time of feeding them that resulted in making the New England barns larger and modifying their design to fit the new requirements.[42]

The sole remaining graphic representation of farm buildings erected before 1660 is the survey of "Ten Hills" made in 1637 showing barns on the Winthrop estate and in nearby Medford. They seem to follow closely the English design. Among other village outbuildings there is what may be a three-story dovecote belonging to John Humfry. In general, barns with porches were entered on the long side, but those with lean-tos must have had doors at each end.[43]

All the barns of which there is a record belonged to the richer inhabitants. At Romney Marsh (Chelsea), the merchant Robert Keayne employed Thomas Joy, an excellent carpenter of Boston, to raise a barn for him in 1640. Joy sawed and prepared the timbers, framed and set up a barn 72 feet by 26 feet and 10 feet in the stud. Like so many yeomen's barns in Essex, it had two porches with great doors to admit carts, but no windows. The barn had a thatched roof, a planked floor, and was sheathed with clapboards. The finished building netted Joy £98 1s. and brought him into prominence as a housewright for the first time. William Codding-

ton's "Large Corne Barne," built in Newport about 1642 at a cost
of £150, burned down in the winter of 1643–44, but within two
years he had two more barns where his men threshed and cleaned
corn for storage. The barn George Wyllys built for his son at
Wethersfield had a "leanto all along one side of it," but no porch;
on the home lot of Phoebe Martin, Samuel Belden, also of Wethersfield, erected a barn that was 28 feet broad and had a lean-to 8
feet wide running its whole length. The written contract stipulated
that "the barne must be built of Good sufficient Timber both frame
and Roofe" and be braced and shingled, and the sides clapboarded. As might be expected, John Pynchon had a large barn, 50
feet by 24 feet, with a large door in the middle of one side and a
lean-to all along the other side with two doors in it. [44]

It was many years before the New Englanders erected buildings to serve as schoolhouses; education was left to the family and
to the minister. Sometime before 1642 Dorchester had a schoolhouse "in good and sufficient repayre"; but what kind of structure
Boston children used in 1645 is not recorded. Ipswich had a grammar school in 1636, but it is doubtful that it was a town school. Not
until the passage of the Massachusetts and Connecticut acts of
1647 requiring all towns of one hundred families to establish Latin
grammar schools do we find records of buildings maintained for
this specific purpose. Located, as a rule, on land granted by the
towns, the first schoolhouses were one-room story-and-a-half
buildings closely resembling the ordinary New England cottage.
The one erected in Dedham in 1649 at public charge measured 18
feet by 15 feet; it had a fireplace 4 feet wide; two windows lighted
the room that had "boarded, feather-edged and rabbitted" walls,
and a planked floor. Stairs led up to the chamber used by the master as a study. Two years later the house was shingled; and the old
frame and catted chimney was replaced by one of brick and stone.
The school building at Watertown measured 22 feet by 14 feet;
anything larger would have been hard to heat during the coldest
weather. [45]

The celebrated brick and stone schools of England, ordinarily
built with funds bequeathed by rich townsmen, could not be imitated in colonial America. However, one signal departure from
common practice did occur in May 1647 when Henry Dunster,
president of Harvard College, contracted with three masons to

construct a one-and-a-half story schoolhouse out of 150 "load of rock stone" brought in carts from Charlestown to Cambridge. The stone walls of the cellar had to be sturdy enough to support masonry one and a half feet thick above the ground, and the gable-ends were to be "wrought up in battlement fashion." The completed structure included a brick chimney, a tile roof, a fireplace ten feet wide on the first floor and one eight and a half feet wide in the room above. Such an elaborate school building proved too costly for any town, including Cambridge, whose inhabitants neither contributed to the work nor helped to maintain it before 1656, when the town bought it for £108 10s. (An additional £30 were paid to the widow of the former president in 1660.) There, for many decades, Elijah Corlet conducted his private grammar school to prepare youths for entry to Harvard College in this substantial stone edifice with crenelated gables that served Cambridge until 1769.[46]

The first warehouses for storing a wide variety of trading goods began to appear with the expansion of commerce by 1639; previously, as inventories reveal, merchants stored grain and many other things in the rooms on the second floors of their dwelling houses, where they were dry and comparatively safe from rodents. Upon receiving a valuble grant of waterfront land at Boston from the selectmen in 1641, Valentine Hill and associates referred to "such warehouses or other houses" as they would erect there. Four years later Thomas Clarke and William Tyng each owned a warehouse on the Cove in that same neighborhood. In 1653 John Johnson, a yeoman of Roxbury, purchased from Joshua Foot, formerly an ironmonger of London, "all that weare house with partitions to Eight Roomes with seller under thirty foote in length and twenty foote in brea[d]th, scituate on the South sid of the Dock in Boston." Plans were being made in 1660 at New London to ship "slitt work for frames for ware houses" to Barbados. Before 1660 all warehouses in Boston and the other New England seaports or trading posts were frame structures built like private houses with the exception that the interiors were arranged differently.[47]

Actually it had been the men of Plymouth who first undertook to adapt the box-frame house with clapboard sheathing for trading houses; two were set up before immigration to Massachusetts Bay began: the first at Aptucxet in 1627 and the other on the Kennebec

River at the site of Augusta the next year. They erected a third in 1633 on the Connecticut River where Windsor now stands. Nothing is known about the Kennebec post, but the beams for the frames of the other two were prepared at Plymouth, and all the materials were assembled there and taken by water to the selected sites.

When in 1628 Isaack de Rasieres went from New Amsterdam to open commercial relations with New Plymouth, he sailed from Buzzards Bay up into the Manomet River, where, in the present town of Bourne, he came upon a "house made of hewn oak planks, called Aptucxet." Here the Pilgrims kept two men the year 'round, and from it they planned to send their newly built shallop to open a trade in wampum with the Indians along the shores of Narragansett Bay. The trading post, now carefully and accurately reconstructed, consists of a story-and-a-half house about 23 feet square. An ell is separated from the house by a brick chimney. The frame and clapboard sheathing are of oak; the shingled roof of pine.[48]

The largest and most ambitiously conceived New England structure of any kind before 1660 was what came to be known as the Old College in Cambridge. Construction commenced in 1638 under the supervision of Nathaniel Eaton. When the General Court dismissed him as head of Harvard College in September 1639, the task of carrying on the work devolved upon Samuel Shepard, who may possibly have had advice from Hugh Peter and Joseph Cook until the new president, Henry Dunster, assumed charge in 1641. Shepard seems to have supervised the carpenter, John Friend of Salem, the plasterer, Richard Harring, "Thomas the smith," and several bricklayers.

Built in the shape of a letter E, when viewed from the front, the college looked like a set of five clapboarded houses joined by party walls. Referring in July 1647 to "the first evill contrivall of the Colledg," President Dunster told the Commissioners of the United Colonies that "there now ensues yearely decayes of the rooff, walls and foundation." It would be unfair to blame the wretched construction solely on John Friend, because he served as chief carpenter only during 1639-40. Any failure with the framing took place under Eaton. The truth is that the undertaking, which took from 1638 to 1645 to complete at a cost between £900 and £1,000, was beyond the resources of the infant colonies—perhaps even more in management and skills than in money. Edward Johnson wrote about 1650 of "the building thought by some to be too gorgeous for a Wilderness, and yet too mean in others apprehensions for a Colledg," especially for those gentlemen who had lived in one of the colleges at Cambridge or Oxford. Nevertheless, this attempt to combine a cluster of familiar forms into one timber structure marked a noble effort, and until 1676, when it could no longer be maintained, the Old College provided a place for the youth of New England to be "brought up in a Collegiate Way of Living."[49]

More than two decades passed before an attempt was made to design and erect another large building in New England, and this time it turned out to be a great success. Expressing a widely felt need, the Boston town meeting voted in March 1649/50 to grant "immunity . . . forever from any rent or toll" to that person who would undertake to build a "howse for the Courts to be kept in." When the rich merchant Robert Keayne died six years later, his voluminous will contained a long section urging the construction of a town house having space for the court, a market, a library, an armory, and a covered place for merchants and traders to walk "on change," all under one roof in a public building. Keayne had in mind an English town hall set upon pillars, such as he might have seen in Essex at the Guildhall of Thaxted or the handsome structures built by John Abell at Ledbury and Leominster in Herefordshire or in many another market town. His designation of the merchant Thomas Broughton and Dr. John Clarke as men "that have good heads in contriving of building" to advise with "some skilfull and ingenious workmen" about drawing up a model for one fabric and his gift of £300 for the project testify both to his thoroughness and seriousness of purpose.[50]

At a town meeting on March 9, 1656/57, a committee of four leading citizens was appointed to procure a model of the town house, estimate the cost, pick the site, and take subscriptions from the inhabitants. A total of 175 persons promised to pay more than £500 in money, goods, grain, provisions, building materials, hauling or lightering, and labor! Meanwhile Messrs. Edward Hutchinson and John Hull arranged that Thomas Joy and Bartholomew Barnard should prepare a model or draft, and after carefully perusing it, they signed a contract on August 1, 1657. This model is the earliest evidence of a large building plan for any of the English colonies in America. The two artisans were to erect a story-and-a-half house measuring 66 feet by 36 feet and resting upon 21 "Pillers of full ten feet high between Pedestall and Capitall," thereby providing ample covered space below for both a market and a merchants' walk.[51]

Thomas Joy served as master builder, and Barnard presumably superintended the brickwork and masonry. In every way Joy, who had lived in Boston from 1640 to 1647, when he moved to Hingham, demonstrated that as an artisan trained to be a carpenter and millwright he had the capacity to advance and become New England's foremost adapter and master builder. He and his partner completed the work on time and to the satisfaction of the authorities, and by March 28, 1659, the selectmen were meeting in the new town building. Save that it was sheathed with clapboards, roofed with shingles, and had exterior closed staircases, the new town house at Boston looked very much like the half-timbered and plastered-daub market, guild, or town halls of old England.[52]

Of all the construction carried on within the still medieval confines of the Forest Vernacular, the most original was the meetinghouse, an edifice used by the people of every town for secular quite as much as for ecclesiastical purposes. The Puritans never intended such a building to be a church. "There is no just ground from scripture," Richard Mather insisted, "to apply such a trope as church to a house for a public assembly." It was John Winthrop, speaking in March 1631/32, who used the term when referring to "the new meeting-house" at Dorchester; by 1636 the figure of speech was a part of common parlance.[53]

The Puritans, it seemed, loathed the word church. The parish churches and cathedrals of England with all of their elaborate exterior and interior decoration and images appeared to the Puritan no better than "remnants of Idolatrie"—Roman edifices that most of them had once been. The pastor of the English congregation at Rotterdam set out this view with great vigor in 1617. "There is not any one place holy, and peculiarly consecrate to the ministration of the Lords supper, as there was of old for sacrifice only at Jerusalem," Francis Johnson wrote. "So as now therefore a place being a generall circumstance that perteyneth to all actions, it hath in this case (as clothes also have) a civill use, commodious and necessarie for people to meet in together, and be kept from injurie and unseasonableness of the weather, etc."[54]

Fixed ideas about what a meetinghouse should be like were not a part of "the cultural baggage" of the first New Englanders, nor did they (as is so often assumed) proceed immediately to raise places of worship. At Salem from 1629 to 1634 the people worshiped in the house built for Governor Endecott, where, presumably, they also assembled in town meeting. Ordinarily the constructing of a meetinghouse followed that of temporary shelters and often after the colonists had moved into permanent houses. In any event, as with almost every feature of "the New England Way," the settlers experimented with and adapted forms of English vernacular building to newly arisen civil and religious needs with materials locally available. When a church was gathered at Lynn in 1631, because the ground that was obtainable was level and subject to high winds, the meetinghouse was made with "steps descending into the Earth" like the cellars dug for primitive shelters.[55]

Most of the first meetinghouses were primitive structures: rectangular in shape, framed and clapboarded, and with thatched roofs. The Bostonians used a small "Mud-wall Meeting House" from 1634 to 1640, when a larger and better-lighted structure replaced it. As with their dwellings, the Puritans had to put up with "Houses of the first Edition," Joshua Scottow pointed out, "without large Chambers or Windows, Ceiled with Cedar or painted with Vermillion," such as those of the second edition later in the century.[56]

During the period 1630–42, forty meetinghouses went up: twenty-nine in the Bay Colony, six in Connecticut and New Haven colonies, and one on Long Island; and in four towns the inhabitants replaced the first structures with new ones. The erection of so many in such a brief period made it possible to try various arrangements within the limits of the familiar frame-clapboard pattern devised by modifying the medieval English domestic building forms. In each one they incorporated such common features as were required for the dual purpose of the building: a room with a pulpit placed at the center of one wall so that everyone might hear the sermon or the speaker at town meeting, and a table below it for either religious use or purposes of the town clerk. As towns multiplied, so did meetinghouses: forty-one more were added between 1651 and 1660, twenty-nine of which were new and twelve, second editions. As before, Massachusetts led in such construction, followed by Connecticut, Long Island, and New Hampshire; Maine acquired one. In 1660 the entire region contained more than eighty meetinghouses.[57]

The growing prosperity and notable improvement in building skills that had produced better and slightly embellished domestic buildings in the older New England towns led to the same advance in meetinghouse construction. The building at Dedham in 1638 measured 38 feet by 20 feet and 12 feet in the stud, but not until 1649 did the town meeting vote to have it "lathed upon the inside, and so daubed and whited over workmanlike."

When Springfield voters authorized the building of a meetinghouse in 1645, they stipulated that every inhabitant (no more than six at a time) should give twenty-three days of work. They also arranged for Thomas Cooper to supervise the work. The building was to be 40 feet by 25 feet, have six glazed windows, three doors, daubed walls, and two turrets—one for a belfry, the other for a watchman. In view of all the free labor, the building was to cost no more than £50. The town of Cambridge gave over repairing its meetinghouse in 1650 and raised a new one 40 feet square, which made such an impression at neighboring Watertown that its selectmen asked John Sherman to "Build a meeting house Like unto Cambridge in all poynts, the Cornish and vane Excepted" for £400 and the seats of the old building. Four months later, after Sherman completed his "pattern" for the building, the town meeting voted to proceed with the work.[58]

A number of meetinghouses put up in the fifties were built under contract and often by a housewright of more than local reputation. The contract for that of Dover in New Hampshire stipulated the use of many of the latest components: six windows, two doors "fitt for such a house," and a tile roof. All walls were planked, the windows glazed, and in the turret hung a bell purchased from Captain Richard Waldron, the leading resident merchant. After living in Seekonk, Dorchester, and Boston, Job Lane settled in Malden. By this time he had gained a reputation as an outstanding house, mill, and bridge builder, so it is no surprise that he was chosen by that town to build its meetinghouse. The contract Lane made with the committee of five on November 11, 1658, called for the construction of "a good strong Artificial [workmanlike] meeting House" 33 feet square, suitably fitted with windows, a pulpit, and seats. The interior was to be lathed all over, "well struck with clay, and uppon it with lyme and hard up to the wall plate." The roof was to be covered with "short shinglings," and atop the edifice a bell was to be hung in a turret 6 feet square "with rayles about it." Job Lane was but one of nine craftsmen whose names we have who labored to provide meetinghouses of seven towns with such improvements; and it was chiefly through their efforts that the meetinghouse style became the colonists' foremost contribution to the Forest Vernacular.[59]

V

The first primitive shelters, followed by permanent houses, of a population ranging from 45,000 to 50,000 New Englanders had required the construction of at least 5,600 to 6,000 dwellings, together with almost as many barns, outbuildings, and structures for special uses: meetinghouses, mills, warehouses, bridges, wharves. All of these were essential for the survival and continuance of the society. Obviously a very great though incalculable number of trees had been felled to provide the strong timbers, heavy planks, boards, clapboards, and shingles that went into all of these buildings. At the same time, all kinds and thicknesses of timber had been cut down and large acreages cleared to prepare ground for cultivation or grazing, and to make into miles of fences and (in smaller amounts) sledges, carts, dugout canoes, crude furniture, barrels, and other casks, tools, and a variety of implements. Much of the smaller stuff taken off cleared land went into cord wood—to supply each New England household with, at the very least, fifteen to twenty cords (a cord measured 4 feet by 4 feet by 8 feet) of firewood each winter entailed further cutting in the adjacent woods. This concerted attack on the forest merely to fill domestic needs brought about a receding of the woodlands immediately surrounding every village and town. The surprising facts are how very soon the timber shortage began and the genuine concern it aroused among the colonists, many of whom had not forgotten the cold winters of Old England and the scarcity of firewood there.[60]

The wisdom of the Puritans in settling in compact agricultural villages encircled by open fields, undivided common lands, and good meadows rather than on widely dispersed farms in an unfriendly country had, indeed, much to commend it for defense, worship, and sociability. Nevertheless, it contributed directly to the disappearance of the trees. The order of the General Court of Massachusetts in 1635 that no dwelling house should be built more than half a mile from the meetinghouse in any new plantation further accelerated the deforesting of such areas. Recognizing that domestic builders had a prior claim on all kinds of lumber, the Court also forbade the export of staves and all "wrought lumber," a measure it repealed in 1640 when economic stress, a lessening of local demands, and increasing production of prepared lumber altered conditions.[61]

Soon after settlement began, the leaders in each town grew alarmed over the waste of wood. In Ipswich so much timber went into buildings and fences that as early as 1635 the town fathers forbade anyone to sell, lend, give away, or convey away "any Timber, sawen or unsawen, riven or unriven," under penalty of forfeiting it, whatever price it commanded; five years later they set a fine of 20 shillings per tree felled on the common land. Plymouth also had become exercised about the destruction of firewood, particularly over the common practice of cutting off just the tops of trees and leaving the remainder standing. The town meeting voted that anyone who felled timber on the commons must also cut up the "bodys" of the trees within a month; a fine of five shillings was set to stop such indiscriminate waste. On Long Island, Southold had been settled only fourteen years when, fearing the total loss of the woods, the town meeting voted that no more trees on the common land could be cut down; this was in 1654. Five years later, however, that same body exempted the making of pipe staves and headings from the ban provided the town's permission was secured. A year before this, the town meeting of Hingham directed the selectmen not to give away "any of the Townes land or Timber or wood." The prevailing attitude in most of the New England communities was expressed at Dorchester in town meeting in

1658: "Timber and fire wood is of great use to the present and alsoe future generations, and therefore to be prudently preserved from wracke and Destroying for the use of the people."[62]

The wasting of forests was far from being confined solely to the neighborhood around each settlement. No sooner did parties of settlers move inland and establish new villages than inhabitants felt the need of a highway to connect them with an older or nearest town. The people of Watertown, desiring a better link with Concord than a path, voted in 1638 that the "Highway leading to Concord shalbe 6 rod broad" (99 feet). In the next year, planning to have the main roads of Massachusetts Bay laid out better for travelers, the General Court ordered every town to choose three men to widen the principal ways "6 or 8 or 10 rods" on all common ground from Newbury to Rowley, to Ipswich, and Salem, on to Boston and from there south to Weymouth, Mount Wollaston, and Hingham—every town, it seems, required a way to Boston or some lesser seaport.[63]

The construction of these coastal highways, which varied from 99 to 165 feet in width, forced the removal of a vast amount of valuable timber and consumed countless hours of labor for which a compelling demand always existed. The width, which seems excessive for a new country, was deemed necessary for farmers who drove their animals to market—room where they could graze the creatures or group them for the night if the journey took more than a day, as well as making it possible for light to penetrate to the road so that the drovers would be aware of any wild animals or hostile Indians. There would be less danger on a wide highway from trees or branches falling across the road during a severe storm as the cart way ran down the center except where it deviated to go 'round a rock, a stump, or a mudhole. It can be noted that the width of the highway south of Boston as far as Braintree meetinghouse was reduced to four rods in 1649. Even a road four rods wide, stretching for fifty or more miles like the coastal highway, necessitated clearing away 400 acres of timber, much of which was bound to go to waste. Watertown may not have been the only place where two men were designated to mark certain trees in the highway with a "W that shall continue for shade," or it may be a lonely instance of concern for the passerby.[64]

In addition to trees cut down for domestic and civic purposes were those felled for what was known as "merchantable timber," an increasingly important source of income: beams, masts, spars, and plank for shipbuilding, and all sizes of lumber, pipe staves, and headings were sawed and split in ever greater quantities after 1640 and shipped to distant markets. To many thinking men of 1660, it had become evident that everywhere New Englanders were rapidly and wastefully depleting their forest resources. All of them knew that westward, northward, and down east great timber reserves still existed, but access to them could be had only in the ships of the wealthy merchant, not by the average settler.

Rarely was the attitude of the emerging Yankee farmer that of "Woodman, spare that tree," but yet, with the memory of England's sad plight still fresh in the minds of many of them, the collective wisdom in most of the towns can be read in the orders aimed at stopping the indiscriminate felling of trees in Massachusetts, Plymouth, and Connecticut; Rhode Island, New Hampshire, and Maine colonists as yet expressed no alarm. Of one thing no doubt existed whatever: colonial concern about the American wilderness originated before 1660 in the town meetings of New England. The militant, unremitting, Puritan attack on the forest had been total, like that on the Pequots in 1637, and it was altogether too successful.

V

"The Famous Infamous Vagrant"
Tom Bell

The printer of the *Boston Evening Post*, Timothy Fleet, commenting in the issue of September 10, 1739, about a letter from Barbados telling of the spectacular successes of an impostor there, concluded dryly: "This Young Blade is supposed (and with great Reason too) to be a Native of this Town . . . and if he has play'd Half of the Pranks, that are imputed to him, we may venture to pronounce that the ENGLISH ROGUE was a meer Idiot, compar'd to him." Westward the course of crime was taking its way, and it must be admitted that its leading adventurer was a Bostonian.

The Age of Enlightenment, the eighteenth century, was also a time of widespread sharp practice. Far more appealing to the popular mind than the renowned *philosophe* was the adventurer. He was a heaven-sent antidote for boredom. Most of his breed were of humble or lower-middle-class origin. Employing their wits and insights into the characters of their victims, they never permitted scruples or morals to hamper their actions. "It need not surprise us," Paul Hazard has written, "that the eternal spirit of adventure takes on the colour of the age. . . . Adventure was a business, a calling, though a calling with a certain glamour about it." Displaying a veneer of culture, a smattering of languages, an excellent memory, and clad in the best of clothes (usually unpaid for), the adventurer would turn up in some great city of Europe or the Brit-

ish Isles. By virtue of his lively discourse, nimble wit, and daring
or effrontery (or both), he passed as a man of distinction, and high
society, as well as low, accepted him. Above all else, he was an
itinerant and, after a brief stay in a place, was gone, ordinarily just
before citizens discovered and revealed him as the charlatan he
was.[1]

After the accession of the first Hanoverian to the throne of England in 1715, the scribblers of Grub Street regaled the reading
public with chap books and longer accounts of the "pranks" of the
great criminals: Jack Sheppard, "prince of escapers," Jonathan
Wilde, informer extraordinary, Dick Turpin, horse-thief, and others. The feats of such scoundrels were romanticized, and in 1728
John Gay in a Newgate pastoral, *The Beggar's Opera*, impressed
the legend upon both his contemporaries and posterity.[2]

Among the lesser adventurers of enlightened England were the
confidence men: common cheats, sharpers, swindlers, charlatans,
humbugs, and impostors of many species. Branded collectively
throughout the British Isles as "rogues and vagabonds," all of them
were *strollers* practicing some form of deceit. One of the best
known of this class claims our attention because he spent two "sabbatical" years perpetrating his deceptions upon credulous colonists
in America. Bampfylde-Moore Carew (1693–1770), the son of a
rector in Devon, has been famous from his time to ours as "The
King of the Beggars." After the manner of many a clergyman's son,
Carew ran away from Tiverton Grammar School and joined the
gypsies and soon became a resourceful rascal, winning wide notoriety as an impostor and simple swindler. Eventually, in 1739,
he was brought to justice in the Court of Quarter Sessions at Exeter where he was convicted as a vagrant and sentenced to be
transported to Maryland for seven years.[3]

When Carew, along with thirty-nine other convicts, arrived in
the Chesapeake, no buyer appeared who wanted a servant who
had no useful trade, let alone a "dogstealer" or "rat-catcher."
When taken ashore, Carew promptly escaped from Captain Froade
of the *Juliana*, but four woodcutters whom he met forced him to
return to the ship. The captain hired a blacksmith to fit him with
an iron collar such as runaway slaves wore and set him to loading
the vessel. Eluding his captors, he encountered two friendly mariners who, though they would not risk the severe penalty for re-

moving his collar, advised him to seek out some amicable Indians who would relieve him of the burden. This he did, and for some time he lived and hunted with the natives. When they moved to the neighborhood of Duck Creek, Carew abandoned his saviors, stole a canoe, and worked his way up the Delaware to New Castle.[4]

Posing as a Quaker or as Jack Moore, a clergyman's son who had been kidnapped and shipped to America, Carew traveled overland from New Castle through Chester to Philadelphia. At Darby "Jack Moore" encountered the great evangelist, the Reverend George Whitefield, who was commencing his first tour of the northern colonies, and found him an easy touch for £3 or £4. In the metropolis Carew, learning by chance that Governor George Thomas of Pennsylvania was a Welshman, extracted money from him by claiming to have a Cambrian wife. The English rogue went as far as Block Island and New London, and at the latter place he imposed upon two aged maidens, the Davies sisters, by telling them that a fortune awaited them at Crediton in Devon. From New London he took passage to England where opportunities to employ his talents and skills were more available and lucrative.[5]

About 1742 Carew (recently elected "King of the Gypsies") was taken up on the quay at Topsham, tried, condemned to servitude, and once more transported to Maryland. This time, well acquainted with the region, he promptly ran away from his master and made his way northward, engaging in the same deceptive practices upon many of the same gullible persons he had duped three years before (so he claimed in his memoirs), though under different names. From Newport he sailed for the Piscataqua, and from Portsmouth traveled southward along the great Eastern Road through Marblehead to Boston. By the time he sailed homeward again, this cheat's unsavory tricks had become a staple of tavern conversations from Maryland to New Hampshire, and his *modus operandi* long remained in Yankee memories. As late as 1818 John Adams, in a letter to Thomas Jefferson stating that he feared the revival of the Jesuits, exclaimed: "Shall we not have swarms of them here, in as many shapes and disguises as ever a king of the gypsies, Bampfylde-Moore Carew himself, assumed?"[6]

The Romany monarch diverted himself during the tedious eastbound crossing from Boston by composing the story of his colonial sojourns of 1739 and 1742. Publication at Exeter in 1745 of the first version of *The Life and Adventures of Bampfylde-Moore Carew, the noted Devonshire Stroller and Dog-stealer, as noted by himself during his passage to America* marked the culmination of the transit of the confidence-man phase of European civilization to the English settlements.

At least as far back as July 2, 1730, the American press had taken notice of colonial strollers who were practicing their arts in the manner of the "King of the Beggars." In the *Pennsylvania Gazette* of that date, its readers learned about "a most successful cheat" named Alexander Cummings who swindled many South Carolinians in a loan-office fraud. Three years later Justice Abijah Savage, Esq., committed John Hill to prison in Boston as "a notorious Impostor" for pretending to have been a prisoner among the Turks who had cut out his tongue and burned his arms with hot irons. Hill, accompanied by Rachel—posing as his wife—took in numerous sympathetic Yankee yokels by showing a parchment setting forth his predicament until the suspicious minister of Abington seized him by the throat and forced Hill "to produce his Tongue or be choaked." Of course the tongue appeared. Nor were these merely isolated instances of low-life.[7]

As soon as newspapers began to circulate in the seaboard cities, the reports about adventurers became the human-interest fodder of the age. A woman called Mary Kemp "but commonly known by the Name of No Mouth'd Moll" was sent to the Bridewell at Boston in November 1736 for "deceiving charitable and well disposed People" all the way from Pennsylvania to Massachusetts. Reputed to be a successful beggar, she was supposed to have a "great deal of Money (some say Pockets full)." At Philadelphia the next year, Thomas Hopkinson, Grand Master of the Lodge, resorted to the columns of Brother Benjamin Franklin's *Gazette* to warn the public about a gang of swindlers who, after they had caused one man's death, were masquerading as Freemasons and fleecing innocent colonists. The very year of Bampfylde-Moore Carew's first arrival on these shores, 1739, the *Boston News-Letter* of November 26 contained a warning about one Robert Palmer (a name used by the infamous Dick Turpin), who pretended to have been left a large estate by John Nash, Esq. of Bristol. It seems he "imposed on several persons and was last seen . . . on the Road Island Road." The

present-day reader of colonial newspapers will be astonished to
discover how promptly the seamy and disreputable features of the
English press won an audience in America.[8]

Actually Bampfylde-Moore Carew achieved only a moderate
success as a cheat; there was very little of the spectacular about his
American tours. By 1739, as Timothy Fleet of the *Boston Evening
Post* realized, a native Bostonian had become the nonpareil. Just
fifty years ago I first came across "the notorious Tom Bell" in the
Pennsylvania Gazette. This uniquely American mountebank fascinated me, and I have pursued him in the colonial newspapers as
resolutely as Inspector Javert did Jean Valjean. Here is my final
report on this Harvard man gone wrong who became, before his
end in 1773, probably the best-known American both on this continent and in the British West Indies.[9]

A son was born to Captain and Joanna Bell on February 18,
1713, in a house on North Street in Boston. The child was named
after his father, a prosperous mariner, shipowner, and shipwright,
and the possessor of considerable real estate in the North End. In
the language of that day the Bell family belonged in the upper rank
of "the middling sort of people" and they aspired to have their
firstborn, at least, move up into the Massachusetts gentry. To assure this the father sent Thomas to "the Free Latin School in
School Street" (the celebrated Boston Latin School) then ably conducted by Mr. Nathaniel Williams. There, from 1723 to 1730, this
very bright lad received from the headmaster and Samuel Dunbar,
the usher, thorough instruction in the grammar, reading, and
speaking of Latin, as well as two years of Greek that included reading in Aesop, Homer, and Hesiod. The curriculum, almost exclusively classical, was aimed to prepare youths for Harvard, and
upon completing it, Thomas Bell's acceptance at the college would
be assured.[10]

Before the lad graduated from the Latin School, his father died
in Nansemond, Virginia, on July 29, 1729. His estate amounted to
£1,535.2.11, of which £1,150 was in real estate and the remainder
in personal property; there were also two household slaves. The
Captain's will provided for £200 to pay for tuition, board, and
room, and incidental expenses at Harvard, but, however, after all
debts were settled, there was almost nothing left. Determined that
the boy should get the education that would make him a gentle-

man, the Widow Bell and her son Daniel petitioned the executors of the estate to be allowed to sell some land on Union Street, and in 1731 Joanna applied to the Selectmen, unsuccessfully, for a licence "to retail strong drink at her house in Ann Street."[11]

Somehow the family managed to find the means to send Tom Bell out to Cambridge in the fall of 1730 where, on September 21, he was ranked 21 in a class of 34—"high among the poor boys," according to C. K. Shipton. That same day Bell complained to President Wadsworth and the tutors that Jabez Richardson, a senior bachelor not in residence, had beaten and abused him; he was awarded ten shillings in damages, though the officials thought it necessary to admonish the freshman privately for his "Saucy behaviour to Sir Richardson." The youth pursued his Latin and other studies without further incident until April 1732, when he accused a classmate, John Wainwright, of stealing his cheese and the faculty exacted payment of "treble the value" of the cheese. In the following December the career of the young Bostonian took a downward course when the faculty ordered him to pay Richard Pateshall ['35] double the value of two bottles of wine he had charged to the latter's account at the Buttery without any order, "either mediate or immediate."[12]

Whether or not Thomas Bell disclosed any dishonest proclivities at the Boston Latin School, there is no question as to the validity of the charges of sauciness and arrogance in Cambridge. Possibly he mistook such behavior as proof of gentility. During most of his freshman and sophomore years he had a chamber in college but ate only a few meals in Commons—odd conduct for an impecunious student. Early in 1733 he got into real trouble. "Whereas Bell has been heretofore convict of theft, for which he was on the second of December last sentenced to a public admonition, the execution of which, upon his earnest Intreaty and Solemn promises of Reformation, was suspended; and whereas he has been Since Found guilty of other acts of theft, particularly of stealing private Letters from Gray's ['34] study, and lies under the strongest suspicion of having stole a cake of chocolate this week from Hunt ['34], and has been guilty of the most notorious, complicated lying, both formerly and more lately, and particularly in the affair of the Chocolate . . . and whereas to all the rest he has added, a Scandalous neglect of his college Exercises, and is now become a disgrace to the Society, and unworthy to be continued a member of it, therefore agreed and ordered that Bell be forthwith expelled [from] the College." This decision by the President and Fellows was read after evening prayers before all of the faculty and students, and Bell was pronounced expelled (he was not present). His name was "erased off the Buttery Tables in the Hall," and all scholars were forbidden to entertain him in their chambers. All this occurred just ten days before he attained his majority.[13]

There were other counts that added to Bell's predicament. Of accumulated fines and charges on the Steward's Ledger amounting to £19.8.1, he was only able to pay £8.10.9, and he had been too prodigal in his endeavors to adopt the tastes of a gentleman: delicate foods, fine wines, and elegant dress. His Boston tailor, Henry Laughton, knowing of his dismissal, sued him on February 24, 1733, immediately after the youth came of age, for £30 long overdue for making him a silk jacket, hose, and other fancy clothing. Laughton won his case in the Inferior Court of Suffolk County, and when Bell appealed to the Superior Court he lost again.[14]

By this time Tom Bell was already "notorious" and *persona non grata* in Boston because of the two lawsuits. He was legally free from family supervision and parietal control by the Harvard authorities, and understandably he decided to leave the Boston neighborhood and take up the familiar career of his father and become a mariner. From what we can learn, his first voyage as a seaman before the mast took him to London, and at some time he crossed to Jamaica where he spent eighteen months working as a storekeeper and clerk. Before 1735 he had also traveled on foot, or at times on horseback, in North Carolina, Maryland, Pennsylvania, and New York.[15]

During these years Bell gradually came to realize that his education—superior to that of all but a very few in English America—could be made the basis for a new and unique calling. The ability to speak Latin fluently and to read Greek was the badge of a gentleman, and so also were his two years at Harvard, his ready wit, facile speech, and parade of upper-class knowledge. In his years at college, he had become aware of and envious of the prestige and privileges of prominent dignitaries and members of the gentry of the several British colonies and filed away in his memory their names and positions. In London he might well have heard and read about Bampfylde-Moore Carew, and by 1736 the example of that impostor may have suggested to him the idea of working out and perfecting techniques for deception and fraud. At any rate, two years later, notices appeared in the press about Tom Bell and, by the next year when the "King of the Gypsies" first came to Maryland, the bad boy from Boston was providing the public up and down the Atlantic seaboard and in the Caribbean with its favorite human-interest news.

The titillating story of the new phase of Tom Bell's career first broke at Williamsburg in the *Virginia Gazette* in July 1738 in the form of an official "Wanted" advertisement:

"This is to give Notice, That on the 2d of this instant July one Thomas Bell, alias Francis Partridge Hutchinson, who was committed to Isle of Wight Goal, for Felony, made his Escape from John Rodway, the Constable of that County, as he was carrying him to Prison. He is a middle-siz'd Man, with his own Hair, and had on a Brown-colour'd Broadcloth Coat, lin'd with Ash colour'd Silk, which has a Rent in the Back of it. This is therefore to desire all Persons to aid and assist in taking up the said Felon, so that he may be brought to Justice, and they shall be satisfy'd for their Trouble by me John Rodway." Nobody, as far as we know, claimed the reward.[16]

The next thing we learn about Thomas Wentworth Bell, "as he called himself," was that he was indicted and convicted by a court in the City of New York on a Friday in mid-November of 1738 "for a common Cheat, for falsely, unlawfully, unjustly, knowingly, fraudulently, and deceitfully, composing, writing, and inventing a false, fictitious, Counterfeit, and invented Letter in the Name of William Bowdoin, Merchant in Boston, directed to Robert Levingston and Peter Levingston, Merchants in New York, thereby to defraud the said Levingstons of £50 Sterling, &c." On the following Tuesday the court ordered the criminal be whipped with "19 Lashes at a Cart's Tail, and . . . drawn through the principal Streets" of Gotham.[17]

James Johnson, almost breathless, rode into Philadelphia on May 30, 1739, with a tale of highway robbery: as he had approached the town, he had encountered a man who had robbed him of £300 in North Carolina and £50 sterling during his journey northward. When Johnson laid hold of him, "The Highwayman beg'd not to be expos'd, and pretended he had marry'd a rich widow in Town, and would immediately refund the Money," if Johnson would go with him to his house. On this pretense he led the way southward out of the city and then escaped to the Lower Ferry, but being pursued and the boat not ready, he fled into the Neck where, some hours later, he was taken up and committed to prison. Two weeks after this incident, Benjamin Franklin assured his readers that the man in prison was "the same, who, under various names" had perpetrated "a great Number of Cheats and Rogueries in this and the Neighbouring Provinces for several years past." Whether Tom was released or escaped from the Philadelphia prison, he seems to have decided to go to the Caribbean until "the heat was off."[18]

As with many another genius, Tom Bell's most flamboyant and historically significant exploit occurred early in life. It was an "ugly affair" that took place at Speightstown, Barbados, late in July 1739. An impressive young man bearing the name of Gilbert Burnet (son of the late Governor Burnet of Massachusetts and the grandson of the famous Bishop Burnet, the historian) had been invited to a Jewish wedding. The groom's father, one Lopez, had made much of the visitor because he spoke Latin, was obviously a gentleman, and had been cast away in a shipwreck. At the nuptial dinner the guest complained of a headache, and Lopez *père* advised him to go into the house and lie down. About half an hour later Lopez *fils,* the bridegroom, and several other young Jews came in, took the gentleman by the collar, stripped him of his finery, and struck him several times in the face, asserting that he had stolen some of their money. All of this the *Boston Evening Post* reported "so surprised him that he could hardly speak." When the news of this beating got out, the Christians of Speightstown were all in an uproar against the Jews; and though the bridegroom could not pin the theft upon the Gentile guest, a justice of the peace bound him over for the Court of Grand Sessions. Tom Bell, for indeed it was he, countered resourcefully by suing young Lopez in two actions of £10,000. Then, the newspaper account continues, "the Leeward People have upon this Occurrence been so irritated that they have raised a Mob and drove all the Jews out of [Speights-] Town and pulled down their Synagogue: The Jews thereupon got together at

Bridge-Town, resolving to have satisfaction; they being generally pretty rich 'tis reported they will be protected by one of the greatest Men in the Island." Thus did the brazen youth touch off the earliest anti-Semitic riot on record in the New World.[19]

The first news of this unsavory affair reached the continental colonies via a letter printed in the *Boston Evening Post* in September 1739, and to it Timothy Fleet appended a shrewd editorial comment. "The young Man above mention'd must certainly be an Imposter, neither of the late Governor Burnet's Sons being in that Part of the World." Another letter from a Bridgetown merchant to "a Gentleman of Distinction" in Salem, dated October 25, 1739, and printed in Fleet's paper on December 10, supplied the public with more sensational news about the Yankee in the Caribbean. The Barbadian frankly conceded that he had ignored timely warnings sent to him from Salem, but not having previously doubted Bell, he had gone security for him. Unfortunately for the merchant, Tom's concealed intention was to make his way to Jamaica where he knew, from an earlier visit, that the take would be much greater. He was in debt to one merchant for £50 and owed £250 to several other people, and so was determined to get off the island and, without any notice, decamped. The writer caught up with him about three miles from Bridgetown before he could board a boat and delivered him to jail. He was wearing costly clothing, including a coat of black velvet in a pocket of which was "an imitation" of the Burnet arms copied from a map of Boston. Since that time, the informant added, he had admitted that he was Thomas Bell and that his mother was alive in the Bay Town.[20]

"In Justice to the World he ought not to live," exclaimed the aroused merchant. He drew bills of exchange in "my Brother Fairfax's name," which he afterward admitted, and also had passed himself off in Virginia as "Mr. Fairfax's brother." In truth, Tom "has the most impudence of any one living; he makes a Joak of all the Evidences that come" from New York and New England against him: "Of his being expelled from the College and tryed also for Robbery in New York, and thinks after all this he has done no Harm."[21]

The court was not impressed by the accused rogue's statements and, toward the end of December 1739, Bell was sentenced at Bridgetown to be whipped and then to stand in the

pillory. It was also ordered that he be "branded with the Letter
on each Cheek, for his Misdemeanours and wicked Actions." Then,
as throughout his life, Tom Bell proved "remarkably Successful in
exciting Compassion in the Ladies," and those of Barbados were
no exception, and by their "assiduous Intercession," some gentle-
dames persuaded Governor Byng to remand the order to brand
him. Had they not prevailed, the public career of the young scoundrel would have ended at once. Even though Bell was not shunned
by many of the Christians, he deemed it advisable to leave Barbados after Governor Byng's "Goodness and Clemency."[22]

No mention of the scapegrace appeared in continental newspapers until in February 1742 the *New-York Weekly Journal* published a letter from Newport informing the public that Tom Bell
had landed there from Barbados in June of the previous year (undoubtedly a second trip to that island). Since then he had been
making a tour of New England and the middle colonies and had
just returned to Rhode Island in the first week of January 1742 in
order to redeem his "fine Cloaths that were attached on account of
the House he hired at Chelsea" in Massachusetts Bay.[23]

This Rogue's Progress had been spectacular, if not rewarding.
Things started off well enough at first and by late June or early July
of 1741 he had arrived at the New Jersey village of Prince Town.
He entered a tavern appropriately and purposely clad in "a dark
parson's frock" and attracted the attention of John Stockton, Esq.,
who mistook the Boston "Stroller" for the Reverend John Rowland
and addressed him by that name. Upon being told of his error, the
Squire, who knew the noted "Newside Presbyterian" pulpit orator
rather well, told Bell that his resemblance to Mr. Rowland was remarkable, and invited him to spend the night at the nearby Stockton mansion. The rascal, ever alert, immediately sensed the opportunity to turn his resemblance to the minister to his advantage.[24]

The next morning he proceeded to Hopewell, then in Hunterdon County. The pulpit there was vacant for the time, and the
ministerial candidate for the Hopewell Presbyterian Church,
Tom's classmate at Harvard, the Reverend John Guild, was away
in New England. Mr. Rowland had preached there once or twice
but was not well known and consequently when the ingenuous impostor introduced himself as Mr. Rowland to a prosperous Presbyterian farmer, the latter invited him to supply the pulpit the next

Sunday and meanwhile to lodge at his house. On the Sabbath, Bell set out for church in a wagon with the ladies, and the master rode alongside on a fine horse. As they neared the meetinghouse, "Mr. Rowland" suddenly remembered that he had left the notes for his sermon behind and, borrowing the horse from his host, galloped back to the farm, rifled a desk in the house, mounted the steed, and sped away to try his luck elsewhere.[25]

The Hopewell incident occurred during the religious revival known as the Great Awakening, which was particularly intense in this part of New Jersey. At this very moment "Hell-Fire Rowland," a terror to all evildoers, was traveling in Pennsylvania, the Lower Counties on the Delaware, and Maryland in the company of the great revivalist William Tennent and two pious laymen, Joshua Anderson and Benjamin Stevens. Upon his return to New Jersey, Mr. Rowland was arrested and charged with robbery.[26]

For the second time in his career "the famous infamous Vagrant" had indulged in a "prank" that led to grave social disturbances after its perpetrator had vanished from the scene. Although the judge of the Court of Oyer and Terminer for Hunterdon County, "with great severity," charged the grand jury to bring in an indictment of Mr. Rowland, after long consideration it hesitated to do so, and, on being reproved and sent back by the court, declined again. Threats by the justices induced the jurymen to give way, and they returned a true bill. At the trial held in Trenton, a clear case was made against the evangelist when several witnesses from Hopewell swore that he had stolen the horse. For the defense, however, Messrs. Tennent, Anderson, and Stevens testified that they had actually heard Mr. Rowland preach in Pennsylvania or Maryland on the very day of the incident at Hopewell. They clearly proved an alibi, and the jury declared the preacher "Not Guilty."[27]

In Hunterdon County the rank and file of the inhabitants were not satisfied with the verdict. The owner of the horse insisted that he was certain the culprit was Rowland; as did many others. Yielding to such public pressure, the Court of Quarter Sessions in August 1741 found bills indicting the three witnesses for the defense "for wilfull and Corrupt Perjury" or false swearing. In November, William Smith, Esq., one of Mr. Tennent's attorneys, procured the removal of the case to the Supreme Court of New

Jersey for the March term of 1742. For some reason the court sent the case to a jury. Just before the trial, as the minister was walking along the street in Trenton, he met a man and his wife who had come from Pennsylvania in response to a dream each had had that Mr. Tennent sorely needed their testimony. They said that they had heard both him and Mr. Rowland preach in their colony at the same time the robbery took place at Hopewell. Two of the ablest lawyers of the middle colonies, John Coxe of New Jersey and John Kinsey from Pennsylvania, joined with William Smith of New York in defending the Reverend William Tennent; their star witness, Judge Stockton, described his meeting with Bell. The combination of legal skills of the eminent counsellors with the supernatural testimony of the couple from Pennsylvania proved too much for the attorney general and his able counsel, and the jury rendered a unanimous verdict of not guilty.[28]

During this colonial *cause célèbre* for which our shabby hero was wholly responsible, he was miles away from Trenton. It may have been during this period that a justice of Sussex County (Delaware) who later was mayor of Philadelphia, committed Bell to prison for forgery. At any rate in January 1742, as we noted earlier, he reached Rhode Island after a tour of eight colonies. After attempting to escape from a Newporter who had lent him money on a previous visit, he was apprehended and clapped into prison.[29]

"The poor Wretch," Zenger's correspondent in the *New-York Weekly Journal* wrote, "is in deplorable Circumstances, and very much in need of Compassion." From prison, Bell sent out on January 14 "a Brief to beg for Relief" after the fashion of English vagabonds and promised "an entire universal Reformation." In it Bell mentioned for the first time "publishing his Journal, on proper Encouragement," assuring that it would be very entertaining. We know, the correspondent declared "that he has already (since he was here last) passed under the Names of Winthrop, De Lancey, Jekyl, Wendell, and Francis Hutchinson: He brought this last from Hudson's River, a little below Albany quite to the Town." The announcing of a book of memoirs placed Thomas Bell with the select company of literate and talented European exploiters of conventional morality and trust.[30]

All literary efforts were put aside when the all-American folk-villain procured his release from the Newport jail—or let himself out (a skill in which he excelled)—and set off once more for the middle colonies. For almost a year the colonial press lost sight of him until on a Thursday night, February 3, 1743, he was apprehended in Philadelphia posing as a "castaway" son of Mr. Livingston of New York and taken before Mayor William Till who happened to remember him as the forger committed to jail in Sussex County some years before. When taxed with that fact, the suspect admitted his identity and was sent to prison for want of security.[31]

The time had come, Benjamin Franklin decided, to advertise the adventurer to the colonial public, and the next issue of the the *Pennsylvania Gazette* (Feb. 10, 1743) contained the most complete spread made so far about Tom Bell. What follows is quoted liberally in order to give the reader something of the flavor of the sensational news and the extent of the coverage achieved by the press.

"He has it seems made it his Business for several Years to travel from Colony to Colony, personating different People, forging Bills of Credit, &c. and frequently pretending Distress, imposed grossly on the charitable and compassionate. He had been in every Colony on the Continent, and in some Parts of the West Indies; and talks familiarly of all Persons of note as if they were of his Acquaintance. In Barbados, he was Thomas Burnet, that rais'd the Persecution against the Jews; in the Jerseys he was the Rowland, that occasion'd so many Prosecutions which ('tis said) he now owns. In Virginia, Maryland, the Carolinas, New York and New England, he has been Fairfax, Wentworth, &c.&c. We hear that a very exact Description of his Person is intended to be published, and dispers'd thro' the Plantations to prevent further impositions."

Supplementary information was supplied to the *New-York Weekly Post Boy* in "a private Letter from Philadelphia." Before reaching the Quaker City, "he imposed the name of Livingston upon one Mathers of Chester, who I suppose has received some Favour from that Family, and this Bell appearing like a Gentleman, persuaded Mathers to let him have two Horses and a Man to bring him up here: *He's now taking his second Degree in our College* [jail], and if he be found fit for the Pillory, doubtless he will receive his Ordination. . . . I was present when he was before the Mayor; he there said he was well acquainted with Books, skill'd in the Law, and Divinity, and believ'd no one could tell him any thing he was not acquainted with before. He rejoic'd in his Sufferings, and hoped to clear himself from all Aspersions."[32]

The details of Bell's escapades having been placed in the public domain by Benjamin Franklin, circulated all over the plantations as the printers avidly followed every exploit and published them in their gazettes. The description of the adventurer promised by "Poor Richard" came out promptly after "THE NOTED AND NOTORIOUS TOM BELL . . . apparel'd compleatly" escaped from the Philadelphia prison on June 16, 1743. It was written by the subsheriff, Thomas Crosdale, who hoped to recapture the man: "He is a slim Fellow, of thin Visage and pale Complexion: Had on when he broke Prison, a dark blue cloth Coat, black silk Jacket, black cloth Breeches, black silk Stockings, new Pumps, with black Steel Buckles; and 'tis supposed he'll wear a new castor Hat or velvet Jockey-Cap and grey Wig; he stole a Great Coat of lightish Brown Colour, with brass Buttons . . . N. B. Among his other Tricks and Villanies he is very dexterous in Picking Locks, having obtain'd his Liberty from Confinement by this Means."[33]

The widespread reports of the "famous American traveller" bore fruit when Tom Bell, after his escape from the Philadelphia prison, made his way from the New Jersey side of the ferry to New York in a "Pettyauger without either Coat or Waistcoat." He went into a public house for breakfast, after which he called for writing materials and "pretended he was going to write some of the most considerable Gentlemen of the City, with whom he was well acquainted, but could not see them until he was well provided with Cloathes to appear in, and said he had lately come from Sea, where he met with Misfortunes and had lost upwards of 600£, and insisted he was the Son of a Gentleman of Fortune at the East End of Long Island." His story was challenged and he was proclaimed to be the felon then publicized in the papers. Picking up his hat and insisting that he had never heard of Tom Bell, he made his way out through the backdoor and, almost as if to make good his word, headed for Long Island. Even there he could not escape notice, for reports from New York located him there on July 11th and again on the 17th, "playing his Pranks" by claiming to be a son of Colonel Floyd and thereby defrauding several persons. Pursued and captured by one of his victims, he was taken as far as the ferry to Manhattan, but again he managed to escape.[34]

The people of New Hampshire were warned on August 22 by the *Boston Post-Boy* "to be on their Guard, for in all probability the famous or rather infamous Tom Bell is upon the line." At least a person exhibiting his characteristics had been seen in the area of Portsmouth and had been swindling some of the trusting natives. "He is very sly in telling us his Name, 'tho to some he said it was Winslow." After breaking open a chest and carrying off considerable money and goods, he left Portsmouth on the 18th, before the notice appeared, well dressed as always. The paper announced that the "Catchpole" was after him.[35]

The next newspaper account of this charlatan's "expedition" in New England appeared in early September. Levying on his Harvard training and experience in New Jersey, he went about impersonating "Hell-Fire Rowland" and preaching in the "New Way" at several places in Plymouth County. But, as the *Boston Evening Post* expressed it, "tis said he has not yet arrived at the Honour and Convenience of an Armour-Bearer" [for the Reverend George Whitefield]. After exhausting all revivalist possibilities in the Old Colony, Bell crossed to Martha's Vineyard and there, "having strip'd an unfortunate Desk of about Ninety Pounds," proceeded by boat to Stonington, Connecticut, and thence eastward to Newport, only to be arrested for "an Old Affair." On this occasion, having the wherewithal to pay, he was not imprisoned and shortly afterward disappeared.[36]

Charlestown, Massachusetts, was Tom Bell's next stop and there, daily from the 15th to the 20th of September, he was at the house of James Hayes, shopkeeper, ostensibly selecting and buying fabrics and having wearing apparel made for him. Then without notice he disappeared, taking with him a piece of chintz, eighteen yards of taffeta, eleven yards of cambric, and five yards of black velvet, the whole valued at £20 lawful money. It turned out that Tom Bell was married to Esther Bell (a cousin?) on October 18, 1743, and it is not unlikely that the theft was effected to ensure that the bride and bridegroom would make a fine appearance at King's Chapel. The brief entry in the "Register of Marriages" merely tells us that the Bells were Anglicans—the most fashionable communion at Boston. It is regrettable that it is the first and last scrap of information we have of Mrs. Thomas Bell.[37]

The couple's honeymoon lasted but a few weeks, for the husband was arrested early in November at Woburn "on suspicion of "Male-Practices." Ann Hayes and her husband Joseph declared that the reason they suspected Bell was "his Cuming to their house and pretending that his name was Benjamin Woodberry, and that he had molases and Bacon Cuming to unlade at Charlestown (which Storey was only a Blind), and the manner of his going a waye with out the knowledge of any bodey in the house." Hannah Baker of Dedham testified that at her house Bell sold her a remnant of taffeta and one of cambric out of an assortment he had with him. Because the accused could not find sureties for £50 to ensure his appearance in December at the next Court of General Sessions, Ezekiel Cheever, Esq., committed him to the common jail at Charlestown.[38]

On the 2nd of December the Middlesex Grand Jury presented "Thomas Bell, lately residing at Charlestown, but now of no certain Town or Place, Labourer," for stealing goods from the house of James Hayes. The defendant promptly pleaded "Not Guilty," but at a hearing held by Judge Cheever that same day, the benedict acknowledged his proper name to be Thomas Bell and that he had told Hannah Hayes "they might call him Mr. Benjamin Woodbridg if they pleased." He further confessed to the theft of the fabrics and to having been taken up at Roxbury or Dorchester shortly after leaving the Baker house at Dedham on suspicion of purloining a wig [to complete his wedding attire?] but he had left his portmanteau as security with Judge Heath. Following the examination he was sent back to prison to await trial.[39]

John Chamberlain, keeper of the jail, seemingly was not aware of "the noted criminal's" propensity for getting "out of confinement," for on the night of December 5, by using a false key, the prisoner escaped. The day after Christmas, Timothy Fleet, in a facetious vein, inserted the following announcement in his paper: "One day last Week, the very famous Mr. Thomas Bell, who lately left his Lodgings in Charlestown [jail] in a pretty abrupt Manner, in order (as he says) to pay a few Visits, and settle his private Affairs was conducted back to his old Cell, where 'tis thought he will be obliged to continue at least three Months, the Court being just over at which he had a case depending, but was unhappily absent."[40]

Upon indictment by the Grand Jury for Middlesex County, Thomas Bell was tried and convicted on March 14, 1744, by the Court, which ordered that on March 21, between two and three o'clock in the afternoon, he "be whip'd at the public Market in

Charlestown, Twenty Stripes upon his Naked Back; that he pay 75 Pounds Lawful Money, being treble Damage, and Cost of Prosecution, and stand committed till Sentence be performed; and that in Case he complies not with the Sentence that he be sold for Three Years."[41]

Dazed and frightened by the severity of his sentence, the rogue now realized that at least in his native Massachusetts his game was up. By March 15 his renowned arrogance had completely disappeared as evidenced by a lachrymose petition he submitted to the justices that took the form of a confession. It was phrased in the semi-pedantic language then deemed polite prose: "Humbly sheweth" that "having for a Term of Years prodigiously mistaken the intrinsic Value of a reasonable Life and madly preferred a Conduct diametrically Opposite which your Petitioner is fully convinced has tended greatly to the Prejudice of Mankind, as likewise to your Petitioner's almost inevitable and irretrievable Ruin both of Soul and Body: That your Petitioner has been for some time sincerely inclined and firmly resolved, by divine assistance, intirely and effectually to revise his past vitious Practices and unreasonable Behavior which your Petitioner in no Sort pretends to justify and your Petitioner does now Promise that nothing for the Future shall influence him to injure the meanest Figure among Mankind in order to save Life Itself or to acquire the greatest Advantages that can belong to it." He also felt obliged to acknowledge that by "the over ruling Hand of Providence" he had often been detected and severely punished for his "enormous Offenses." All this notwithstanding, the petitioner "has remained inflexible and unreformed and (by Divine Direction almost hardened in his impenitency)" and therefore "justly exposed to the judgment of Almighty God and the Condemnation of Men. But now under the greatest concern for and lively sense of his condition, and concerned only with integrity and universal reform," he, Thomas Bell, is amending his life and manners. Humbly and heartily he begs forgiveness of the court, and implores the justices to correct their just sentence "by some alleviating Ingredients or favorable Alterations."[42]

Our sources do not tell us whether Bell's sentence was carried out or whether once more he foiled justice by breaking out of jail. That he was not sold as a servant is clear, because in mid-July two

Philadelphia papers reported that "the famous Sharper" had been apprehended in the city and sent to jail for an "old Misdemeanour." A fickle colonial public no longer thought of Tom Bell as aught but a disreputable loser, and the man himself was desperately descending deeper into crime.[43] In August, five men and six women were imprisoned at Philadelphia as accomplices in counterfeiting ill-engraved New-York bills of credit in a log house in a remote part of the Jerseys. "The Discovery was first made by Tom Bell, to whom one of the Gang apply'd, endeavouring to engage him to be concerned with them, and to assist [with his skill as a penman acquired at Boston Latin School] in Signing their Bills." There was small honor in this thief who won his freedom by squealing.[44]

Soon Bell changed locality by moving southward where the winter was milder and his practices were far less well known. His remaining "fans" in New England and the middle colonies had to follow his rogueries in the newspapers, whose printers allotted space to accounts of his travels in the South and to the Caribbean.[45]

At Charleston on February 11, 1745, Lewis Timothy, using information gleaned from northern papers, announced in his *South-Carolina Gazette* that "The Publick are desired to take Notice, that the famous Tom Bell, alias Burnet, alias Rowland, alias Fairfax, alias Wentworth, alias Livingston, alias Rip Van Dam, &c. &c. &c., is arrived among us. He was in Town last Week, and had taken upon him the Name of Nathaniel Butler, and has signed a Paper Y. P. Randolph. He is now in the Country, where beware of him—He is a slim Fellow, thin visaged, appears like a Gentleman, talks of all Persons of Note as if intimately acquainted with them, and changes his Name and Cloathes very often."[46]

A week later the readers of the *Gazette* learned that "the celebrated Mr. Thomas Bell . . . so well known thro' all the Colonies and most of the Islands by his Actions for several Years past," has been committed to the Workhouse after the Honorable Edmond Atkin, Esq., of the Council had detected him on the road to Goose Creek posing as a native, Nathaniel Moore. And, continued Timothy, "as a great Concourse of People daily flock to see him, no doubt the further Execution of his Schemes in this Province will be hereby defeated." Since no crime was alleged against him, the

"Vagabond" was released in a few days, and shortly he disappeared from the Low Country.[47]

During the next six months the Caribbees were the scene of Bell's activities. Then, in the fall of 1745, he landed at New Bern in North Carolina and went up the Neuse River on his way northward. Back in December of 1744, using the name of Robert Middleton, he had forged a letter of credit from Col. Andrew Meade on John Campbell of Edenton and received a sum of money from the Colonel. As soon as Campbell heard of Bell's return to Bertie County, he hastened to insert a warning in the nearest newspaper, the *Virginia Gazette*, published in Williamsburg, against "an incorrigible Offender" known to be in the area. Five pistoles and all costs were promised to anyone who had the rascal arrested and jailed either in Williamsburg or Edenton within six months. An added note stated that Bell had brazenly told some Neuse River gentlemen about September 23 "that he should make free with their Names as it might suit his Convenience in travelling Northerly" as Captain Randall from Havana. Accordingly, trading on the names of the Tarheel gentry, Thomas Bell made his way back to his northern haunts where he hoped that during his long absence he had been forgotten.[48]

One might reasonably suppose that the newspaper owners would find the accounts of Bell's arrests, convictions, and escapes too stereotyped after a time to give much space to them. There was often, however, some incident that varied enough in details to warrant reporting it.[49] Such a one was the "Act of Villainy" related by Captain Charles Dingee, master and owner of a river boat at New Castle. On August 8, 1746, a person calling himself Lloyd and claiming that he owned a ship boarded Dingee's boat about three miles up the Delaware. A gentleman in appearance and behavior, he was left on board at the wharf when the vessel reached Philadelphia. "Contrary to the Principles he endeavored to counterfeit," he slipped away with three Holland shirts, some handkerchiefs, stockings, and, above all, "one pair of scarlet Breeches" belonging to Captain Dingee. The skipper offered a reward of £2 to anyone jailing the crook, but it is extremely doubtful if Dingee ever recovered his cherished red pants, for within a month news came from New York that "the noted Tom Bell" had enlisted as a soldier in a New Jersey company under the command of Captain Stevens and gone with it to Albany.[50]

Campaigning against Indians and Frenchmen was not Tom Bell's forte, and he carefully calculated the price and judged it too low. In September 1747 he again turned up at New York and on the 27th was thrown in jail. Inasmuch as neither the Supreme Court nor the Court of Quarter Sessions had received any charges against him during the first week of November, he was discharged after apologizing for affronting a magistrate and told that he was to leave the city within twenty-four hours or be sent to the House of Correction as a vagrant.[51]

Time was manifestly running out on Thomas Bell. His obsequious confessions and promises of reform had proved hollow and spurious. When he was imprisoned at Newark in East New Jersey in September of the following year "on suspicion of coining Money," one printer hoped that this time he would not escape "as he is an Old Offender." Nevertheless he did get free somehow, for by December he was once again in serious trouble in his native Boston.[52]

Timothy Fleet broke the exciting news in his *Boston Evening Post,* January 30, 1749, that "on Thursday last Sentence of death was pass'd on a *notorious old Offender,* who has been as great an instrument of Fraud, Oppression and Injustice, as has been known, perhaps, in any Age of the World; notwithstanding which the Tryal lasted several Weeks, and many of the Jury strove for a favorable Verdict. We hear that the Execution is respited until the 31st of March 1750, but it is said, that without his Majesty's special Order, a Pardon cannot be granted." Whether the King pardoned Tom Bell we shall probably never know, but somehow he escaped the gallows and prison as well.[53]

For more than six months after Tom Bell managed to elude the death sentence, no reports appeared in the papers as to his whereabouts. Then it was learned that he had arrived in New York on the last Thursday of August 1749, but he prudently departed the next morning or he would have been arrested—possibly sent back to Boston. This hasty visit inspired Franklin's former associate, James Parker, to animadvert in his *New-York Weekly Post Boy* upon the doings of this very American scoundrel: "Tis somewhat surprizing that this Fellow should so many years continue rambling here in America playing dishonest Pranks; for as he is now pretty well known throughout the Country, his Impositions pass only upon innocent and unwary People; tho' several honest Persons have suffered at Times, because they have been suspected to be like him: He must certainly now be a Pain to himself, as well as an Injury to Society; and as he seems almost incurable in that Sort of Life, it is not unlikely but his Craft may one Day fail him, and he have the Reward in a Triple Tree [the gallows]." [54]

Touched to the quick by James Parker's hard thrust, the now unmasked "villain and imposter" ventured to reply in a long letter headed "North America, August 30th, 1749," which he sent to a rival printer. Henry De Foreest ran it in the *New-York Evening Post* of September 4th. This *apologia pro vita sua* (as the author would have said) was penned in the florid semi-didactical style that passed popularly as elegant writing. " 'Tis impossible that I should be the Derision of all the discerning World, for my Impieties and Absurdities," the swindler began. Maintaining that "the Compiler and Printer" of the *Weekly Post Boy* was the tool his enemies used "in order ('rem, remquecumque rem') . . . to render me contemptible and impious in the Eyes of the Publick." This "news Writer" spreads spurious charges and lies about me "like false Dice," and delivers them "with big sounding Words without any Truth . . . fit to swell the Throat of a Bully, and keep up the Huzza's of a Bear-Garden Quarrel." [55]

At Manhattan "I never actually offended," insisted the Yankee Rambler. In it "I have bought and paid some score Pounds Worth of Goods and Merchandize for the Space of 12 Years [since 1738], have sail'd as a Fore-mastman out of this Port, enlisted as a Soldier in the neighbouring Province, Taught School in almost every Government in America, survey'd Land, &c. &c." In spite of all this I am treated as if I were some "hideous monster." Why should the magistrates deny the liberty of their city to "persons of my romantic and comical Humour and Reputation." The newswriters assert him to be "incurable of my Sort of Life" when they ought to give him one more chance. [56]

On and on Tom Bell argued his case (over three columns of the paper) in a vein reminiscent of his sophistical pleas for mercy and promises of reformation that he addressed to the justices in Massachusetts. Prosecutions, imprisonments, censures, and reproaches are for slaves and incorrigible offenders only, but are "entirely lost upon persons of Education and ingenious Natures" like himself. Almost tearfully Tom Bell pleaded with James Parker to retract his charges. "I am not wholly lost to all Sense of Honour or Honesty . . . but what with a little indulgence, . . . and a generous Pardon, I may be easily reclaim'd." Then at long last, descending from what Parker accurately branded "Matchless bombast" into abject banality, our author advertised in a postscript that if any gentleman, owner, or commander of a trading vessel bound for any North American port wants "an able and expert Pilot or Sailor" he may apply to *"The slandered and unworthy Person of Tom Bell."* [57]

Learning that "the noted Vagrant, Tom Bell is still lurking near this City, peradventure to have the better opportunity to deceive those he intends next to visit," James Parker scored the final point in the *New-York Post Boy* of September 11th: "This Fellow has had a *large Swing* over this Continent, as well as in some of the West India Islands, for those several Years; and had it not been for some of [John] *Gay's Monkeys,* who flock round him in every Place, *grinning applause* to his *redundant Chattering,* and thereby supporting him in his unparallel'd impudence; 'Tis needless to say more, than that publishing him in the News-Papers, is only to put honest People on their Guard, and prevent their being imposed on by the Wiles of a Man, chiefly protected by Qualities of Craft obnoxious to Christian Society." [58]

Colonial legal machinery was so clumsily designed and distances so great that the clever Tom Bell usually found ways to circumvent the first and take advantage of the latter. When he could not do this, he often took a "sabbatical" leave from the road and served society for a while as a schoolmaster. No doubt he operated on the theory that because not much talent and training were expected in a practitioner of pedagogy—and as a student at Harvard for two years he had both—he could live incognito until the storm blew over. The respite from his death sentence expired on March 31, 1750, by which time he was wielding the ferule upon and teaching about gerunds to the sons of tobacco planters in Tidewater Virginia.

Schoolmaster Bell did not reappear in public until July 17, 1752, this time in Williamsburg. Reporting this event in his *Virginia Gazette,* William Hunter observed: "As his former Character and romantick Life have made a great Noise in every American colony, 'twill doubtless be a Satisfaction to all who have any Knowledge of him, to hear in what Manner he has lived.—He has resided in Hanover County in this Colony near Two years past, in the private Station of a School-master, and has during this Time, behaved himself with Justice, Sobriety, and good Manners, of which he has produced a Certificate, figur'd by the principal Gentlemen of that County. By this behaviour, and his future conduct, he hopes to wipe off the Odium that his former Manner of Life had fix'd on him, and thereby to approve himself a useful Member of Society." [59]

The famous colonial impostor was now forty years old and had been "going straight" for some time when he appeared in Williamsburg. Nothing occurred during the four days he stayed there to give anyone the idea that he would revert to his old ways. When William Hunter noted the pedagogue's return to Hanover he expressed his faith in Bell by the following comments: "If this Man, after the many Pranks he has play'd, is really Sincere in his Professions of Reformation, and his Intentions of living an honest and industrious Life, it may perhaps be a Surprise to many, who are apt to say, *Can the Aetheopian change his Skin, or the Leopard change his Spots*, then also may he do good that has been accustomed to do evil." [60]

It had been borne in on Tom Bell for some time that the reports of his bizarre doings found a wide reception among the reading public. By talking glibly with William Hunter, he succeeded in persuading him to join in a project that had been conceived on Rhode Island ten years earlier. Like every great adventurer of the eighteenth century, Bell stood ready to play a trump card, and this with the substantial assistance of one of the leading American printers. Within three weeks of the visit to Williamsburg, the *Virginia Gazette* conveyed the good tidings to provincial lovers of the sensational in the form of a letter grandiloquently phrased from the schoolmaster, "To the Public": "As I have for some Years past been fully convinc'd of the pernicious Consequences of an unsettled and rambling Life, and of giving the Reins to exorbitant Passions and unlawful Desires, which have been too predominant with me, I am now determined to spend the Remainder of my Days in a close Application to some reputable Business, wherein I may render myself a useful Member of Society, and acquire a Subsistance suitable to my Genius and Education. And as I am at a Loss how to lay a Foundation for my future Liveli-

hood, I humbly propose to lay before the Public a faithful Narrative of my Travels, and Adventures, for upwards of twenty Years
past [since 1730]; which I would publish, by the Subscription of
such Gentlemen as will be pleased generously to favour the Undertaking; which I have been encouraged to hope a great many
will do. . . . And as I propose this Advantage to myself, so I am
hopeful, that the Relation of many Transactions of my Life will be
not only an agreeable Amusement, but also, in some Degree,
useful to others; as therein the World may see by what unjustifi-
able Steps I often proceeded, and learn thereby to avoid those
Snares and Temptations, by which I have often been entangled
and overcome; and which will prove equally hurtful to others; who
are so unwise as to follow the like extravagant Courses."[61]

This amazing confession of wrong-doing was followed by proposals for printing "THE TRAVELS AND ADVENTURES of the famous
TOM BELL . . . Together with a Brief Account of his Birth, Parent-
age, Education, &c." The conditions for subscription are interesting. As soon as the subscription lists were full, copy would go to
Hunter's press in Williamsburg, and books (bound at 15 shillings
or stitched at 10 shillings) were promised on or before September
1, 1753. Bell invited "a voluntary" deposit of a piece of eight from
"such Subscribers only as incline to expedite the Undertaking and
relieve the Author's present Necessity." Otherwise payments were
to be made at the time of delivery. The notice also stated that
upwards of forty gentlemen had signed for copies of the book the
day the subscription was opened.[62]

A "Paper of Proposals" was dispatched immediately by water to
New York, whence the news was spread by the newspapers
throughout the middle and northern colonies and to South Carolina and the West Indies. Christopher Saur of Germantown notified his readers of the *Pensylvanische Berichte* that Bell had "settled down" in Charleston early in September and was holding a
school in the Low Country. The printer may have been mistaken
about the place or else Bell had returned to the Old Dominion; a
hurricane of September 15, which destroyed much of the southern
city, may have rendered it impossible to write his book there and,
especially, to solicit subscriptions. In any event Saur informed his
large German audience all about the would-be author and drew a
Christian lesson from his career: "A fine example. If a Godless one

is converted from his evil ways, then everyone should no longer remember his many sins."[63]

There are almost no clues to Tom Bell's pursuits from September 1752 to July 17, 1754, when, once again, he arrived in Charleston seeking additional subscriptions for his "Memoirs." He had Lewis Timothy append new proposals for an enlarged and more elaborate "faithful proposal" of his "Travels and Adventures for Upwards of Twenty Years past," to the promises of reform as published in the *Virginia Gazette* and other northern newspapers. Lewis Timothy professed himself convinced that the author's account was genuine and that he would not omit anything significant relating to his enterprises and adventures that he would compile from his own records. Furthermore, Timothy declared, "the various tragical and comicall Turnes of his Fortune, may perhaps vie with any History of the kind ever yet made public," (Casanova's *Memoirs* would not be published until late in the century).[64]

From the new "conditions" published in the *South Carolina Gazette*, it seems clear that the former vagabond was sincere about his memoirs. The large octavo volume would be printed at Williamsburg and, as soon as the subscriptions were completed, he guaranteed to send the books from there "in a short Time after Publication." The price in South-Carolina currency would be £5, "if bound," or £3 "if only stitch'd." Probably because of the tardiness of northerners in subscribing, Bell stressed that no more copies would be printed than engaged for, "And the Money not to be paid 'till they are delivered," excepting only generous and humane South Carolinians who will "consider the great Charge and Expence of travelling some Hundreds of Miles in order to procure Subscriptions." The author personally took subscriptions at the Sign of the Bear "up the Path," and lists were also open at a number of other public houses in Charleston. When he announced in the newspaper his intention to leave the southern metropolis before August 23, 1754, the new and proper Thomas Bell politely returned his "humble acknowledgments to the Worshipful the Magistrates" for their "Indulgence, &c.," hoping that during his short stay in their city "my behaviour will intitle me to their further indulgence, if not procure some degree of their favourable opinion."[65]

A careful search by the writer has not turned up a single copy of Bell's "History." It seems most likely that, despite the author's policy of no payment before delivery, his long-standing reputation as a cheat and impostor deterred interested men from subscribing, and as a result, in Williamsburg, Hunter declined to proceed with the printing. No advertisement appeared in any colonial newspaper nor any mention of the work.

This supposition is borne out by the statement of Benjamin Mecum who was publisher of the *Antigua Gazette* in December 1754 when Tom Bell turned up at St. John's for a few days before going on to another port. The traveler appeared to be in a poor state of health but thought that the mild winter in the West Indies would enable him to recover and, at the same time, pursue his publishing design. Mecum was a nephew of Benjamin Franklin and a former apprentice of James Parker of the *New-York Weekly Post Boy*, and he undoubtedly knew Bell's history. The printer obviously had reservations about his visitor's chances of seeing the book in print, for he editorialized shrewdly for his readers: "As a reputed Liar is seldom believed when he speaks the Truth, so a Man who has once acquired the Name of a Villain, is liable to the Imputation of Crimes and Enormities, which he never committed; that this is Bell's case, admits of not more Doubt than the Falsity of common Report, and may be as easily proved as that he has been a Rogue and an Impostor. Who can believe that he never swore a prophane oath, or was drunk? yet, this he affirms; Who can credit, that he never took advantage of, or debauched, Virgin-Innocence? he declares it for Truth: Who can imagine that he never stole a Horse? he clears himself in that Particular. . . ." After rehearsing Bell's career as a traveler, sailor, and teacher, Mecum told of his also practising physic, and pleading law. "*Since, therefore, it appears that his Proceedings do not partake of Tragedy, he seems to deserve the Appelation of an Universal Comedian.*"[66]

The daring young man with the Harvard education whose escapades had for twenty years inspired curiosity, amazement, fear, and envy—and also imitation—throughout the continental and insular provinces was now, at the age of forty-three, being subjected to public derision by the very printers whose newspaper stories had made his feats household tales. Not only had he become too well known but age was forcing him to abandon his youthful style for cheating naïve colonists, and for Americans of every class, he was a legendary figure. In 1755, for example, the eminent Daniel Dulany, the Younger, asserted in a newsletter describing one Belchior, a minister turned con-man in Maryland, that "there was hardly ever so great a rascal—he had imposed upon almost as many people as Tom Bell." Seven years later Benjamin Mecom, then printing in Boston, issued A Serious & Comicall imaginary Dialogue between the famous Dr. Seth Hudson and the noted Joshua How . . . containing . . . a Touch on TOM BELL."⁶⁷

Between 1755 and 1771, as the legend grew, the actual doings of "the famous American traveller," with one exception, were no longer reported: in the New-York Journal of June 10, 1771, John Holt printed an extract of a letter brought from Kingston, Jamaica, by Captain Engler, telling of the conviction in court on March 11 of "Thomas Bell, the late Chief Mate of a Sloop belonging to Montego Bay, for piratically robbing and plundering a Spanish Schooner near Cuba, of 14,000 Dollars. This was the Person who was concerned with Captain Yarr, in October last [1770]," in an act of piracy. The two men were convicted of the crime by the testimony of twelve of their ship's crew, whom they turned adrift at sea in the Spanish vessel after they had robbed it—"without Water or Provision" they fortunately arrived at Kingston a few days later. Condemned to die, the captain and Bell took the sacrament together on March 24; the next day Yarr was hanged, and on April 21 "the notorious Tom Bell" ended his life ignominiously and, as the letter stated, died "penitently."⁶⁸

In the ten years following his expulsion from Harvard College in 1732, Tom Bell became the most widely known individual in all English America before the advent of the revolutionary generation. More people knew what he looked like, how well he dressed, how he imposed upon and cheated innocent colonials of all walks of life, where he had been in North America and the Islands, and all about his unparalleled impudence, fine education, elegant diction, and supreme arrogance than they knew about any other American including Cotton Mather, Benjamin Franklin (as yet), or any royal official or politician. His was the first and one of the most durable and lasting of the reputations to be created by the American newspaper press, whose printers found him ever good copy.

This peripatetic also provides irrefutable evidence of the surprising amount of intercolonial travel in the years after 1730 and of life on the colonial roads and tavern life, not merely by rogues and vagabonds, but by men and women on legitimate business or pleasure.

One can even say that this earliest confidence and bunco artist and his lesser imitators contributed in their peculiar fashion to an emerging sense of American unity.[69]

VI

Philosophy Put To Use:
Voluntary Associations for
Propagating the Enlightenment
in Philadelphia,
1727–1776 *

In the latter part of 1760, Benjamin Franklin put a question, now famous, to Polly Stevenson: *"What signifies Philosophy that does not apply to some Use?"* Throughout the eighteenth century no belief was more tenaciously held by philosophers and men of capacity and public spirit generally, whether on the continent of Europe, in the British Isles, or in the English colonies in America. David Hume and other enlightened Scots gave this idea the central place in their commonsense thinking: and in *Candide* (1759), Voltaire had his Venetian nobleman say disparagingly of an eighty-one-volume set of the proceedings of a learned society that it would have had some value "if a single one of the authors of all that rubbish had invented even the art of making pins; but in all those books there is nothing but vain systems and not a single useful thing."[1]

On New Year's Day 1768, in the room of the Union Library Company on Chestnut Street, Charles Thomson exhorted his fellow members of the American Society held at Philadelphia for Promoting Useful Knowledge to make a fateful decision. "Knowledge is of little Use," he told them, "when confined to mere Speculation; But when speculative Truths are reduced to Practice,

*This essay was reprinted from the *Pennsylvania Magazine of History and Biography*, Volume CI, Number 1 (January 1977), 70–88, with the permission of the editor.

when Theories grounded upon Experiments, are applied to common Purposes of Life; and when, by these, Agriculture is improved, Trade enlarged, and the Arts of Living made more easy and comfortable, and, of Course, the Increase and Happiness of Mankind promoted; Knowledge then becomes really useful." One could search a long time before uncovering a better statement of the ideals and objectives of the philosophers of the Enlightenment as they came to be understood and carried out in America.[2]

II

Let us begin, as the colonial Philadelphians did, in Western Europe, in the British Isles in particular. In England and her colonies, these years embraced an age of freedom quite as much as an age of faith, and many men believed that the cause of mankind could be served in secular and humanitarian, as well as in spiritual ways—no one cried *Écrassez l'infâme*, even about the Church of England. The destructive and nostalgic attitudes so conspicuous among the French found no acceptance in the Old British Empire. Englishmen, Scots, and Americans were optimists, confident that progress was a fact and that, by putting philosophy to use, they could achieve widespread happiness.

We perceive at once that the corpus of ideas, theories, programs, and schemes that constituted the Enlightenment in Europe and England was never exported to the colonies whole and intact; nor did its votaries ever intend that it should be so transmitted. What was dispatched westward was deftly and unobtrusively selected with the greatest care to represent and buttress a point of view, which was all the more influential because, for the most part, the men who held it did so unconsciously.

Searching questions occur to us now: by whom were the Enlightened ideas and programs chosen, and how did they make the selection? What auxiliary ideas and programs were sent off at the same time? How and when did they cross the ocean? And of particular concern to us is whether the Enlightenment was being exported in any organized or institutional forms, or was the transit to America merely a series of random occurrences or the chance work of a few individuals?

Answers to all of these queries cannot be given here, but it is possible to sketch some of the conditions that governed the selection and transmission of certain features of the Enlightenment across the Atlantic. In fact, a realization of the potency of the all-pervading effect of these situations and influences is prerequisite to our understanding of the course of the Enlightenment in America. That this has been either ignored or not recognized in the histories is probably because, at the time, it was so self-evident.

It must be remembered that nearly every colonist's view of England and the rest of Europe and, conversely, the outlook that most Englishmen had on America were images filtered through, as it were, Dissenters' spectacles, the lenses of which had been ground in the British Isles. Why this is so is quite clear: communications were almost entirely carried on by English merchants, the majority of whom were Dissenters. These merchants owned and freighted most of the ships that transported people, books, news, letters, ideas, and sentiments to colonial seaports, and especially, after 1720, to Philadelphia; they also controlled most of the news coming in from their counterparts in the New World. The volume and influence of the transatlantic correspondence have never been assessed, but we do know that it was of prime importance in the struggle over the attempt to create an Anglican Episcopate in America and in several other instances.

The merchants of London, Bristol, and Glasgow were not only Protestant but radically so, and therefore zealously anti-Roman Catholic; they were also overwhelmingly Nonconformist and completely out of sympathy with the Established Church; and not a few of them had experienced religious persecution. It was only natural therefore that they became closely allied with the ministers and spiritual leaders of the Nonconformist sects. Although loyal Hanoverians to a man, they subscribed to many of the republican ideas of the Commonwealthmen of the seventeenth century. Denied access to the universities and public office because they were not Anglicans, the lay and clerical members of the several denominations found outlets for their talents and physical energies in the industrial, educational, and humanitarian, and scientific activities that contributed so notably to the brilliance of the Enlightenment in Great Britain.

The signal achievement of the middle-class Protestant Dissenters was their amazing success in putting philosophy to use by forming and employing voluntary associations of every imaginable kind to mobilize the many like-minded people of ability for the attainment of desirable ends. This was a phenomenon that originated in the seventeenth century—notably with the Royal Society of London—and it became a pronounced tendency among Englishmen and Scotsmen. Unable to attain their goals through Parliament or the Church of England, the Dissenters gave the voluntary association its most extensive, varied, and successful development.

Dissent itself had given rise to the largest and most powerful private organizations and institutions: the nonconformist Presbyterian, Independent, and Baptist churches, which, by 1702, had loosely united as the Body of the Three Denominations. Thirty years later, prominent laymen formed the Protestant Dissenting Deputies to seek repeal of the Test and Corporation Acts and look after the "Civil Affairs" of the Nonconformists.[3]

Far surpassing these dissenting bodies in their impact on the American colonies and dwarfing the Anglican Society for the Propagation of the Gospel in Foreign Parts and the Society for Promoting Christian Knowledge, was the older and far-flung Society of Friends. The skill displayed by George Fox in organizing his Quaker followers for proselytizing and ensuring their solidarity in the face of all detractors has often been commented upon. His genius in foreseeing the possibilities of using yearly, quarterly, monthly, weekly, and special meetings to foster trade and commerce among Friends, and also for humanitarian enterprises, not merely among the widely scattered Quakers on both sides of the Atlantic but for any human beings needing assistance, has been understressed. Here was a mighty voluntary association that, in addition to ministering to the religious, business, educational, and social needs of the membership, contributed to the Enlightenment one of its best known symbols—the Good Quaker. In accounting for the remarkable success of this institution, primary importance must be given to the fact that its transoceanic strongholds were in London and Philadelphia, and that many branch offices had been set up in Britain and the colonies.

The great majority of Englishmen who traded with America and those who corresponded with the provincials upon religious or cultural matters were either members of one of the Three Denominations or of the Society of Friends. They lived and had their being in an atmosphere of Dissent and knowingly screened, if they did not dictate, the ideas and programs that traversed the ocean in both directions. When colonials, even those who were communicants of the Established Church, voyaged to the British Isles, the society in which they mingled was far more often composed of Dissenters and unbelievers than of Anglicans. Consequently they acquired most of their impressions of the mother country, along with selected elements of the Enlightenment, in a Nonconformist environment. From the English Dissenters also, they learned how to apply the principle of association, and not a few of them returned to America inspired and determined to apply organization to the promotion of useful knowledge.

III

Benjamin Franklin made his first visit to London when he was in his late teens; there this alert, impressionable, and highly intelligent young man widened his horizon in countless ways through observations and discussions—we are staggered as we read of what he learned of life in England and of his associations and meetings with people of high and low degree. Ultimately, however, after eighteen months in the metropolis, and now twenty years old, he sailed back to Philadelphia in July 1726 because, as he always contended, "Pensilvania is my Darling."[4]

Franklin's accounts of his early life reveal that his mind was ever brimming over with projects for self-improvement and human betterment. No device for carrying out useful and worthwhile adjuncts to civilized living was equal to the *club*. As a lad of fourteen Franklin had possessed and read with profit the classic English treatment on voluntary associations, Daniel Defoe's *An Essay upon Projects* (1697), and he had drawn heavily upon that work of the sturdy Dissenter for his "Silence Dogood" essays, which were printed in his brother's *New-England Courant*. When he was seventy-two, he recalled that Cotton Mather's *Essays To Do Good*

(1710) had influenced him greatly, but there can be no doubt that his London experience had contributed most to his desire for improvement of himself and his fellow man.

In the fall of 1727, a year after his return to Philadelphia, Franklin tells us that he "form'd most of my ingenious Acquaintance into a Club for mutual Improvement, which we call'd the Junto." (Shortly a Marylander would define clubs as "consisting of Knots of Men rightly sorted.") The celebrated admissions questions that Franklin drew up for the Junto he had appropriated almost without change from the "Rules of a Society, which met once a week, for their Improvement of useful Knowledge and for the promoting of Truth and Christian Charity." He had come upon them in his copy of *A Collection of Several Pieces* by John Locke published at London in 1720. This, it appears, is how the magic, motivating phrase "promoting useful knowledge" was incorporated into the Philadelphia vocabulary. Like some other famous phrases, this was a legacy from the great Mr. Locke.[5]

The very fact that the youthful printer was able to assemble a group of ten talented and serious craftsmen and tradesmen to form "the best School of Philosophy, Morals and Politics that then existed in the Province" indicates that by 1730 Philadelphia was becoming "the outstanding, probably the first, example in the Western World of a culture resting on a broadly popular base." Its 11,500 inhabitants of several nationalities lived and worked together as one free and happy people, tolerant, self-governing, prospering, full of hope, and persuaded by their own experiences and what they saw about them of the possibility of achieving a better future for humanity. Progress for them was not an idea, not a dream; it was a visible reality. Later in the century Saint-Just would declare in France that "Happiness is a new thing in the world," but in the English colonies it had long prevailed and nowhere as widely among people of all ranks as in the city on the Delaware.[6]

In order to facilitate finding information to answer questions raised at meetings, the members of the Junto agreed to pool their personal books, but very soon they found the collection inadequate. Drawing upon what he had learned about subscription book clubs in England, such as the one Philip Doddridge belonged to at

Kibworth in Leicestershire in 1725, Franklin proceeded to set on foot his "first Project of a public Nature" when, on July 1, 1731, he and several cronies of the Junto "erected a common Library."[7] This was a significant step in the cultural history of American society. It signaled the emergence of a great colonial leader, an American *philosophe* who has rightly been designated the finest flower of the Enlightenment. In addition, the occasion marked the beginning of the most successful application of the principal of voluntary organization in America during the half-century before independence. Of this act, the subscribers were well aware, and they emphasized it when they adopted the motto: *Communiter Bona Profundere Deum Est* ("To pour forth benefits for the Common Good is divine").[8]

Through the good offices of Joseph Breintnall in 1731, the officers of the Library Company formed a memorable relationship with Peter Collinson, rich Quaker merchant of London, collector of exotic plants, Fellow of the Royal Society, and a correspondent of many of the leading scientists of continental Europe. Until his death in 1768, Collinson advised the Library Company about acquisitions and personally handled its purchases from the booksellers of London. When *A Catalogue of Books Belonging to the Library Company of Philadelphia* came from Franklin's press in 1741, the 375 titles listed represented the principal interests of the Enlightenment, particularly natural and speculative philosophy and standard works of history and literature. By 1770 readers could choose from 2,033 books, most of them in English.[9]

From time to time nonmembers were permitted the free use of the books. About 1730 Joseph Breintnall had introduced John Bartram by letter to Peter Collinson as "a very proper person" to procure American botanical specimens for him; and the Londoner soon countered by urging that Bartram be made an honorary member of the Library Company. This was done in 1743. In 1774 and throughout the War for Independence, the delegates to the Continental Congress were accorded borrowing privileges, as were the members of the Constitutional Convention in 1787. With its fine collections located in Carpenters' Hall, this institution richly deserves to be better known, in Edwin Wolf's happy phrase, as "the Library of the Revolution."[10]

The usefulness of the Library Company led to the founding of other subscription libraries in both the city and reaches of Philadelphia. The proliferation of such organizations was largely accomplished by Quakers. In 1742 the Junior Library Company of Philadelphia was established. Twenty-seven craftsmen and tradesmen, most of them Friends, organized the Union Library Company of Philadelphia in 1747. This group initially housed its books in a room on Second Street. Its catalogue of 1754 lists 317 titles. Two more libraries were established in 1757: the Amicable and the Association. At this time, as many as four hundred Philadelphians, predominantly middle class, the majority of them Quakers, subscribed to and frequented the four libraries. During the next ten years the last three consolidated and were, in 1769, incorporated into the Library Company. Between Trenton and Darby and as far westward as Lancaster, at least ten other subscription libraries were founded before the end of the colonial period.[11]

Writing in 1774 about the collections brought together under the aegis of the Library Company, the Reverend Jacob Duché observed: "To this library I have free access by favour of my friend the merchant, who is one of the company. You would be astonished . . . at the general taste for books, which prevails among all orders and ranks of people in this city. The librarian assured me, that for one person of distinction and fortune, there were twenty tradesmen that frequented this library." And again he wrote: "Literary accomplishments here meet with deserved applause. But such is the prevailing taste for books of every kind, that almost every man is a reader. . . ." In putting philosophy to use through its books, the Library Company of Philadelphia was probably the most successful institution of its kind in the known world.[12]

In their eagerness to improve "the Minds of Men," certain bold spirits of the Library Company proposed to go well beyond the forming of a subscription library confined to the printed word. They aspired to render "useful Sciences more cheap and easy of access," for not only the members but for others who were devotees of natural philosophy. The success of the Library Company of Philadelphia with extracurricular activities was due in large part to the fact that its leaders maintained important connections throughout the transatlantic groups of Dissenters, including the Society of Friends, and they also had deep interests in natural philosophy.

But the combining of a subscription library, a museum, and a laboratory in one institution, and the opening of it to the public at large for reading, writing, listening, viewing, and experimenting without charge was *the American thing*.[13]

The credit for initiating and promoting the scientific activities of the institution must go to Joseph Breintnall, the oldest and sole merchant member of the Junto. Before 1730 he had made contributions, later deemed worthy of publication in the *Philosophical Transactions of the Royal Society*. As the first secretary of the Library Company, he began to correspond regularly with Friend Peter Collinson about book purchases and also about his own scientific investigations, and he also served as a link between James Logan, Quaker bookman and virtuoso, but *no joiner*, and the Library Company. He encouraged the botanical work of John Bartram and it was he who procured introductions through the London merchant for such Anglican medical members as Thomas and Phineas Bond to M. Jussieu of the Jardin des Plantes at Paris and Johann Friedrich Gronovius in Leiden. In his capacity as secretary of the Library Company, he wrote the letter welcoming John and Thomas Penn to their province, May 11, 1733, and soliciting proprietary "Countenance and Protection" of the infant organization, which already closely resembled the academies so much the vogue in the cities of Europe. In acknowledging receipt of the secretary's paper on the aurora borealis, July 11, 1738, Peter Collinson intimated that he considered Joseph Breintnall the *principal founder* of the Library Company: "All thy observations and schemes relating to it, are an instance of thy zeal for promoting the good of mankind, and deserves the greatest commendation from all that are well-wishers to so noble a design." Breintnall helped guide the institution until his death in 1746.[14]

During the decade of the thirties, Benjamin Franklin's interests centered primarily on books, the promotion of self-improvement, and getting ahead in the printing business. Sometime after 1740, awakened by his Quaker colleague and business associate Joseph Breintnall, the printer learned the ABC's of natural philosophy and won admission to the "natural history circle" presided over by Peter Collinson, whose friendship would prove of such vital importance for his later career.

The arrival in April 1738 of an air pump and "other curious instruments of great Use in the Study of Natural Knowledge," a present from John Penn, prompted the directors to make two far-reaching decisions. The first was to order a local cabinetmaker to fabricate a handsome "cabinet" with a glass front. The instruments, in their beautiful case, were displayed to a better advantage when the institution's collections were moved in April 1740 to two rooms on the second floor of the recently completed west wing of the State House. A little over a month later, the directors permitted the use of the air pump and the "outer room" for a course of philosophical and experimental lectures by Isaac Greenwood, formerly a professor of mathematics at Harvard College.[15]

The acquisition of more apparatus in 1741 and 1745, the issuing of the first printed catalogue of books in 1741, and the chartering of the library in 1742 encouraged more frequent use of the "outer room" for lectures and as a public museum. Esquimaux instruments and utensils brought back from the Arctic voyage of the *Argo*, financed by Maryland, Philadelphia, and Boston merchants, were presented to the museum by the North West Company in 1753 and 1754—probably at the instance of Benjamin Franklin. Matthew Clarkson, mapmaker, surveyor, and later clerk of the Philadelphia Contributionship, bargained shrewdly and successfully in 1761 for a free life membership in exchange for the donation of a cabinet of fossils that had been collected by the late Quaker merchant Samuel Hazard.[16]

In Robert Aitkin's freshly launched *Pennsylvania Magazine* of February 1775, Tom Paine singled out for praise "the cabinet of Fossils" at the Library Company, together with "several species of earth, clay, sand &c., with some account of each, and where brought from." These curiosities, he insisted were valuable not alone because they entertained the scientist but also because they revealed the hidden parts of nature to the potter, the glassmaker, and allied artisans. Mining ought to be developed, said he prophetically, for the bowels of America have as yet been "only slightly enquired into," and he went on to describe a set of "Borers" or drills to be used in sounding the earth. Here, indeed, was a concrete proposal to put philosophy to use, the objective of the Enlightenment, and it was suggested by a visit of this newcomer to view the "cabinet" at the Library Company.[17]

The first Philadelphian to propose gathering some of the "most

ingenious and curious men" of the English colonies into an "academy or society" for the "study of natural secrets, arts, and syan-
ces," and to "communicate . . . discoveries freely" was John Bartram. He seems to have been inspired by the Royal Society of
London, about which he had read in the *Philosophical Transactions,* and letters from correspondents in England. Early in 1739
he broached the matter in a letter to Peter Collinson, who replied
on July 10 that considering the "infancy of your colony," the plan
was not then feasible.[18]

John Bartram did not give up his scheme; he merely bided his
time. When he first met Cadwallader Colden in 1742, it is not unlikely that he brought up the project. If so, he would have learned
that about the year 1728 the New Yorker had urged Dr. William
Douglass to found "a Virtuoso Society" at Boston, and possibly,
too, that Paul Dudley, F.R.S., had been requested in 1737 to form
a branch of the Royal Society in that town. The internationally
known botanist and the locally prominent printer had come to
know each other well by 1742 when Franklin revealed his nascent
interest in science by supporting, in the *Pennsylvania Gazette,* the
creation of a travel fund for his friend Bartram. The botanist's
dream of an academy appealed to Franklin, who wrote out "our
Proposals" and published them May 14, 1743, in a broadside entitled *A Proposal for Promoting Useful Knowledge among the British Plantations in America.*[19]

The American Philosophical Society was organized in 1744 by
Franklin, Bartram, and Dr. Thomas Bond with great hopes and
enthusiastic encouragement from both sides of the Atlantic. By
1745, however, it was languishing, and John Bartram wrote dis-
consolately to Colden on April 17: "most of our members in Philadelphia *embraces other amusements that bears a greater sway in
their minds.*" Two years later, the organization had ceased to
exist.[20]

The demise of this learned body by 1746 has been attributed
with considerable reason to the outbreak of the war with France
and the failure of some local members to give over their pursuit of
wealth and the spending of it in luxurious living and display. In addition, there seems to be an even weightier explanation: the objectives set forth in "our Proposals" appealed chiefly to the devotees
of natural history, but, as Collinson pointed out to Bartram in

1739, the Philadelphia members of the Library Company were not merely well equipped with apparatus but they had been providing the space, the proper conditions, and the leadership expected of an academy. Further, it is more than probable that after 1742 advanced age or illness kept Joseph Breintnall away from his good work, which would explain why so prominent a natural philosopher was not listed among the members of the short-lived American Philosophical Society. After Breintnall's death in April 1746, Benjamin Franklin succeeded his friend as secretary of the Library Company: henceforth, for the first time, he corresponded with Collinson and received from him regularly "the earliest Accounts of every new European Improvement in Agriculture and the Arts, and every Philosophical Discovery." The present writer is convinced that the quick shift of interest away from natural history to experimental science (or physics) at this time by the secretary and his associates was because those "other amusements," of which John Bartram complained, bore "a greater sway in their minds." In any event, the compelling desire to have an intercolonial philosophical society vanished as the group became totally absorbed in learning about the new electrical fire.[21]

Benjamin Franklin, along with certain members and several fellow workers, made the Library unquestionably the center of organized scientific activities in America between May 1744 and June 1757, when the secretary departed for England. There was no institution to compare with it at home or abroad. The "outer room" at the State House became the scene of intense activity as these Philadelphians turned it into a workshop, a kind of laboratory for philosophical and practical experimenting. Dr. Archibald Spencer introduced these eager men to the latest scientific fad by performing (albeit somewhat imperfectly Franklin later thought) some electrical experiments during a course of lectures, which were twice given at the State House between May and July 1744.[22]

The most pronounced impetus for undertaking the "Philadelphia Experiments," however, came from overseas when Peter Collinson, in the second half of 1745, sent to the Library Company a copy of *The London Magazine* containing an account of recent German electrical discoveries, together with a "Glass Tube" and directions for using it. Franklin was fascinated as he read the article, and soon three other shareholders in the Library—Thomas Hopkinson, Philip Syng, and Lewis Evans—and the exceptionally perceptive Baptist minister, Ebenezer Kinnersley, were devoting most of their leisure time to devising and constructing the instruments and equipment needed for performing the experiments, which they proceeded to carry on from 1746 to 1749; the gift to the library of a complete, new electrical apparatus by the Proprietors in 1747 greatly aided them in their undertaking.[23]

Results of this work, which Benjamin Franklin dutifully reported to Collinson in a series of epoch-making letters, reveal that, among other things, the astounding success was achieved by an organized or group effort of members of the Library Company in the "outer room." Its theorist and chief experimenter informed the London philosopher on March 28, 1747, that "he was never before engaged in any study that so totally engrossed my attention and my time as this has lately done"; and again, "If there is no other Use discovrd of Electricity . . . it may *help to make a vain Man humble.*" In a later letter, dated April 29, 1749, he admitted to Collinson that the members of his group were still "Chagrin'd a little that We have hitherto been able to discover Nothing in the Way of Use to Mankind." After 1756, however, he was able to put philosophy to use with his lightning rod.[24]

At London in 1751, Edward Cave published, with a preface by the eminent and influential Friend, Dr. John Fothergill, *Experiments and Observations on Electricity, Made at Philadelphia in America, By Mr. Benjamin Franklin, and Communicated in several Letters to Mr. P. Collinson of London F.R.S.* By initiating this project, Peter Collinson introduced both the City of Philadelphia and its leading citizen to the enlightened public, and within a few years both place and man had become famous throughout the Western World.

At the same time the news was being spread up and down the American colonies. Samuel Dömjen, a Transylvanian who had studied at Oxford and whom Franklin had instructed in the performing of electrical experiments, set out on a profitable lecture tour to Maryland, Virginia, North Carolina, and Charleston in 1748; later he went on to Havana and Jamaica. Two members of the library coterie, Ebenezer Kinnersley, in 1749 and 1753, and

Lewis Evans in 1751, further popularized the electrical discoveries from Boston to South Carolina and in the West Indies.[25]

The excitement generated in Europe and America by the "Philadelphia Experiments" ought no longer to be permitted to obscure other philosophical undertakings at the Library Company during this same period or the numerous practical benefits to colonial society that originated in the "outer room." We do not contend that the directors carried out a set program, but we do insist that given the excellent collection of books, the use of the room, the apparatus and cabinet in it, the opportunity thus afforded for men with ideas to meet and try them out experimentally with like-minded persons, the ready access to the Philadelphia press, and the organizing talents and potentials for leadership in the membership, the Library Company actually did meet all of the requirements of a learned academy. Furthermore, the guiding spirits in many of the enterprises were Quakers, the conspicuous exception being Benjamin Franklin.

When the principal activities at the Library Company are merely listed, they are still impressive. Of prime importance were the contributions of the members to the geography and cartography of America. In 1741, Lewis Evans and Thomas Godfrey determined the meridian of their city, and on *A General Map of the Middle British Colonies,* 1755, Evans made Philadelphia the first meridian of America because of its central position and also "because it far excels in the Progress of Letters, mechanic Arts, and the public Spirit of the Inhabitants." On maps of 1749, 1750, and 1752, as well as that of 1755, he incorporated much new data about the western country that he had gathered on a journey to Onondaga in 1743 in the company of John Bartram and Conrad Weiser, the Indian agent. Requests by many Englishmen to read Bartram's "plain yet sensible" journal of this trip impelled Collinson to arrange for its publication at London in 1751. A year later, the surveyor general of Pennsylvania, Nicholas Scull (one of the first shareholders), joined with George Heap in bringing out *A Map of Philadelphia and Parts adjacent,* and in 1759 appeared Scull's *Map of the Improved Part of Pennsylvania,* which contained a more accurate delineation of the western settlements than that of Evans.[26]

Benjamin Franklin, the printer-turned-philosopher, made an outstanding contribution to useful knowledge when, in 1744, he issued his account of the Pennsylvania fireplace, which he had devised in the winter of 1739–40. The final production of the stoves was truly a cooperative venture by members of the Library Company, for Robert Grace built the fireplaces at his Warwick Furnace in Chester County from drawings made by Lewis Evans, who also sold the stoves in the city and its environs. Franklin introduced Peter Kalm, the Swedish naturalist, to the Library Company, and Lewis Evans prepared a very good map of the middle colonies in 1750 for that scientist's book of travels. The idea that American population doubled every two-and-a-half decades was being discussed all over the colonies during the forties, and it is difficult not to believe that Franklin brought up the subject with the directors and other members of the Library, as well as in the Junto, before 1751 when he wrote *Observations on the Increase of Mankind, Peopling of Countries, etc.*, which was first published at Boston in 1755. The essay was a *tour de force* in the field of population studies; it also adumbrated the theory of the influence of the frontier in American development.[27]

At mid-century, as Quaker ideals and humanitarian impulses were more and more being directed toward realities, members of the Library Company figured prominently in the founding of a new association of profound significance to the city and the colonies. The Anglican Dr. Thomas Bond first conceived the plan to erect a hospital for the sick and insane poor similar to those of London and Paris, but, because his scheme was "a novelty in America," he failed to get the necessary backing. Benjamin Franklin, a genius in political maneuvering, proposed to the Assembly that it should appropriate £2,000 if a like amount could be raised by voluntary contributions. Through the effective use of the press, over £2,750 was donated; the contributors immediately appointed managers to carry out the project. A fellow member of the Library Company, the master builder Samuel Rhoads, drew up the plans that provided the institution with the most modern type of hospital then known. Although Pennsylvanians of all faiths were among the 693 "Contributors" (1751–76), Friends made up a large majority of them.[28]

Back in 1749, Dr. Thomas Bond and a group of Anglicans

joined with Benjamin Franklin, several Presbyterians, and two Quakers as trustees in founding an academy, which opened in 1751; in 1754 it became the nonsectarian College of Philadelphia. At this time the Society of Friends did not support higher education; its leading members rightly feared that the new institution would become a political pawn. Franklin saw to it that the "Mathematical School" of the Academy had the necessary scientific instruments and "a middling apparatus for experimental philosophy." The first master was a mathematician, Theophilus Grew, and the presence of Ebenezer Kinnersley, Francis Alison, and William Smith on the faculty meant that natural philosophy would be stressed at the College. In a short time, this institution, under Provost Smith's urging, began to displace the Library Company by offering lectures and providing space for experimenting with its apparatus. The opening of the Medical School in 1765 brought about increased emphasis on the biological aspects of science and a linking of the school to the Pennsylvania Hospital through Dr. Thomas Bond's clinical lectures in 1766. [29]

In June 1757, Benjamin Franklin left Philadelphia for England, and before the outbreak of the War for Independence he was back in his "darling" Pennsylvania for only two years, 1762–64. The great work of organizing the citizenry for putting philosophy to use had been accomplished by this time; his co-workers carried on some remarkably successful activities right down to 1776. In like fashion, Franklin's Quaker associates had pretty well made their contributions to these same undertakings by using with great effect the overseas facilities and personnel of the marvelously organized Society of Friends. The alliance of meetinghouse and printing house—of the Friends and Franklin—had produced not only the Library Company of Philadelphia but a number of offshoots from it, voluntary associations of immeasurable benefit to the public. The Library Company had now, however, served its purpose as a laboratory and museum; after 1769 the responsibility for organized scientific work in Philadelphia was assumed by the new American Philosophical Society founded that year. [30]

So well had the Philadelphians managed to put philosophy to use that their achievements were recognized and acclaimed in England and throughout the Western World. From Naples on May 18, 1776, Abbé Ferdinando Gagliani advised his friend Mme.

d'Épinay, who was planning a change of residence in Paris: "The epoch has arrived of the total collapse of Europe, and of transmigration to America. This is not a jest: . . . I have said it, announced it, preached it for more than twenty years, and I have constantly seen my prophecies come to pass. Therefore, do not buy your house in the Chaussée d'Antin; you must buy it in Philadelphia."[31]

VII

The New England Town:
A Way of Life *

The process by which New England was colonized was unique. Where settlement in other colonies was achieved by individuals or sporadic groups, unrestrained by prior design, in New England the community and social ideal, controlled and directed by provincial authority, was present from the beginning. Groups of religiously and socially like-minded families organized themselves into a "church and town," acquired a tract of land about six miles square, settled thereon according to definite rules, and proceeded to work out for themselves an orderly agricultural community. The New England town was a carefully "planned society." The system, as originally conceived and administered by the Massachusetts General Court, prevailed throughout the province of Massachusetts Bay, in its offshoots, Connecticut and New Hampshire, and its colony of Maine.[1]

Since the church and town were coeval and coextensive, the community formed around meetinghouse and village green. Land was divided by lot into village plots fronting on the green, and outlying portions for farming, grazing, and wood lots. Other tracts were set aside for the common use of all, and still others were held for future needs. This arrangement of home lots in one locality

*This essay introduced the J. Franklin Jameson Lectures at Brown University in 1939, and was published in the *Proceedings* of the American Antiquarian Society in April 1946. It is reprinted with the permission of the Director.

made for a compact agricultural village, effective for religious, political, and social purposes. Community control of the distribution of land implied also control over admission of new residents. As a result the town never grew too large; instead it threw off spores, which in time developed into new towns.

At this point I desire to make it clear that I am writing only of the *rural* town, not of such seaboard centers as Boston and Newport, which I have elsewhere described as cities.[2] Moreover, I will confine my remarks strictly to the period from 1740 to 1776, because in those years the New England town ceased to be a frontier community preoccupied with the pioneer problems of clearing, shelter, subsistence, and defense, and attained its fullest and most effective development. Earlier years reveal the rawness of youth; later years exhibit the decadence of age. It is the rural New England in which grew up the embattled farmers that I wish to examine.

By 1740 the New England town was no longer a social experiment, but a fully developed, functioning institution. One hundred years of interaction between physical and man-made environments had produced a noteworthy fusion. Town and nature exhibited a harmonious adjustment, the visible result of which attracted the attention of all English travelers. "In the best cultivated parts" of New England, reported one of them, "you would not in travelling through the country, know from its appearance, that you were far from home. The face of the country, has in general a cultivated, inclosed, and chearful prospect; the farmhouses are well and substantially built, and stand thick. . . ."[3] Over all prevailed the evidence of careful planning—planning in the lay-out of the land, the arrangement of houses, and the adjustment of the whole to terrain and landscape. The socio-religious community ideal of the inhabitants found expression in an air of permanence and order that had the rigid logic of Calvin's *Institutes*. Stripped of all nonessentials, the New England town had achieved a complete social functionalism. Like its severely simple farmhouses it rejected superfluous ornament, but with the same tendency for cheerful asymmetry, it could afford to ignore unimportant eccentricities. It attained the beauty of entire usefulness. This beauty was unconscious and austere, to be sure, but it was the only successfully blended beauty of natural and man-made environment that America has ever known. A solid unity among the people produced this synthesis and ensured its success.

As this town was not so much an area as a way of life, an understanding of it must be sought in its people. In the four decades preceding the American Revolution, the population of New England more than doubled, increasing from about 325,000 to 675,000. Much of this gain was the result of natural increase of the original English stock or of immigration from England, but this accounts for only about 70 percent of the inhabitants. The remaining 30 percent was composed of Scots, Scotch-Irish, Celtic Irish, Germans, and other folk. These were the people who shaped town life and were, in turn, shaped and molded by it. New Englanders were not merely old Englanders transplanted; they were a new breed; they were Yankees.[4]

Wherever they lived, in village or on isolated farm, these Yankees made the family their fundamental and enduring association. The family tie derived its strength as much from its economic and social as from its moral and religious significance. Not only was it a means for increasing and perpetuating the race; it was also the principal unit of agricultural, industrial, and business life, and the basic agency of social intercourse. It guaranteed the permanence, stability, continuity, and orderly development of the community. With ten or twelve children, several relatives, the hired-men, servants, and slaves, all living together, frequently under the same roof, the colonial family was very large. The larger the family, the more hands to aid in cultivating the soil and doing the chores, the more members for the church, the more citizens for the town.[5]

It must not be assumed, however, that members of one family occupied the same plane as those of every other family. No, the town exhibited a definite hierarchy of social classes. Look about eighteenth-century New England where you will, you will scarcely discover "the pure crystal of democracy." Class distinctions sprang full-blown into being with the division of lands when the town was founded. The few whose position in life or whose wealth seemed to demand it had received more land than those lower in the social scale or less favored by fortune. Privileged families were thus able from the beginning to adopt and maintain a style of comparative luxury. The minister occupied a position of social, as well as religious, leadership, and with his unencumbered homestead, a salary of £60 to £100, exemption from taxation, and free firewood, he could often live much better than the rank and file of his congregation. Madam Estabrook, consort of a Connecticut parson, boasted among other possessions three silk and three crepe gowns, with petticoats to match, three riding hoods, two bonnets, two other hoods, eleven night caps, twenty-one aprons, nine pairs of gloves, and two fans. Such frippery is never found in the inventories of the "middling or inferior sorts."[6]

Frequently the innkeeper, the keeper of the general store, the lawyer, and the physician were also ranged with the select few who composed the upper crust of town society. But most aristocrats belonged to the squirearchy, the country gentry who constituted the leading agricultural, political, religious, and cultural interest of the New England countryside. In the settled areas of Massachusetts, Rhode Island, Connecticut, and New Hampshire, wrote the author of American Husbandry in 1775, are "many considerable landed estates, upon which the owners live much in the style of country gentlemen in England." Many of these large properties were divided into two portions; one cultivated for the squire's use under direction of an "overseer"; the remainder let to tenants under lease. This writer's criterion for a place in the squirearchy was an income of £200 to £500 per year from an "estate," of which rents often formed an important part. More frequently, perhaps, tenants being scarce, the country gentleman and members of his family farmed his whole estate.[7] Supplementing his rents, or the income from sale of his produce, were profits from lumbering, a saw mill, or the sale of frontier lands, which he almost inevitably possessed.

Travelers discovered New England gentry living in a "genteel, hospitable, and agreeable manner." Southern Massachusetts came to be known as the "land of the Leonards," where in baronial splendor dwelt that great family of ironmasters. Daniel Leonard, Harvard College, 1760, who would

> Scrawl every moment he could spare
> From cards and barbers and the fair,

dazzled all New England, including John Adams, in 1767 when he and his bride toured the countryside in their elegant coach before settling down at Taunton Green in their spacious mansion with its deer park.[8] Squire Edmund Quincy of Braintree, Massachusetts, owned a fine "manor house" built in 1716, surrounded by stable, barn, sheds, cider-mill, fish pond, garden, wood-lot, orchard, and broad-acred farm. The interior was beautifully furnished in "the newest fashion" and the best of taste, chiefly with mahogany and fine plate. In 1775, in preparation for the marriage of his daughter Dorothy to John Hancock, the walls of the mansion were hung with imported Chinese wallpapers. In fact, as one observer noted, New England "country gentlemen are enabled to purchase whatever they want from abroad."[9]

Members of the gentry made their influence felt in every sphere of town life. At church they sat in the choice pews, and more frequently than not controlled the elders and deacons. They ran town meetings, either by continuous occupancy of the key offices of moderator and selectman, or by securing the election of townsfolk who saw things their way. As town proprietors, their voice in the regulation and disposition of undivided lands was decisive. Thus did the New England squires see to it that the town thought as they thought, and acted as they would have it act. They were men of ability, accorded respect, if not deference, by the rest of the town. They controlled town life because they possessed capacity and education, as well as wealth.

"The most considerable part of the whole province" of New England, however, was the yeomanry or, as contemporaries defined them, the independent, industrious, freeholding farmers. In contrast with the genteel existence amid comfortable surroundings enjoyed by the gentry, the farmer and his family lived plainly and frugally in a severe, unpainted frame farmhouse. Beyond prime necessities, he might possess a Bible, a musket, a looking glass, and perhaps a piece of carpet. His food and drink were plain and probably monotonous. Throughout rural New England, flour was a luxury indulged in chiefly by the upper class. Thomas Hutchinson told King George in 1774 that the average New Englander preferred coarse bread made of rye or corn to the wheaten variety, and as late as 1800 corn meal remained the yeoman's staple food. Fresh meat appeared rarely on his table. Each autumn a hog was slaughtered and salted down, to be served regularly at future meals. This constant use of salted meat generated the insatiable thirst for which the Yankee became notorious, and which he variously treated, according to his purse or the season, with generous applications of beer, cider, or rum. He thereby supplied his parson with inspiration for countless homilies on the evils of "baneful intemperance."[10]

Soil and climate conspired to force the New England farmer to struggle incessantly for a bare living. Economy and thrift were essential to survival, and the Yankee farmer rationalized his necessity into a creed and made a cardinal virtue of unremitting physical labor. Yet even the most industrious could not avoid a touch of wistfulness. Theodore Foster of Brookfield, after entering on the margin of his almanac for August 18, 1767, "Began to plow behind the orchard," broke into limping verse:[11]

> Thus our work is nevr Done;
> But uneasily it goes on,
> At least for my own part,
> This I say from my Heart.

Although basically a farming community, each New England town had its little coterie of artisans and craftsmen. The blacksmith, the wheelwright, the weaver, the cabinetmaker, and other workers, needed to render the community self-sufficient, may also be classed with the yeoman stock. Of course every farmer had to be a jack-of-all trades, and many tiny industries were conducted in the home, but the importance of village artisans and craftsmen is brought home to us when a village house burned down in 1747 at Preston, Connecticut. One Avery, a weaver, and his family of fifteen lost their looms and £400 "worth of Cloath of their own Manufacturing." Certain towns, moreover, boasted craftsmen whose mysteries served the needs of the surrounding regions or even the whole colony. Caleb and Robert Barker of Hanover, Massachusetts, cast bells up to two thousand weight at two shillings per pound, which graced the belfries of meetinghouses all over New England. Lexington rejoiced in its "noated clock-maker," Robert Mulliken, while Colchester, Connecticut, took as much pride in the fine furniture made by Pierpont Bacon as did Little Rest in Rhode Island in the beautiful silver produced by Samuel Casey.[12]

Toward the close of the colonial period the rural population of New England, especially that part of it which enjoyed contact by water or highway with Boston, experienced a rise in its standard of living. Tea and coffee came into general use after 1750, because farmers were producing a surplus "sufficient to buy such foreign luxuries as . . . [were] necessary to make life pass comfortably." The best-informed observer of the period believed the yeomanry to be in general "a very happy people."[13]

This "middling sort," then as now, formed the backbone of the village community. They were a hard-working and self-respecting lot. To their economic and social superiors they cheerfully accorded due respect but no subservience. Socially as well as politically the yeomen were sure of themselves. They were as jealous of their rights as any colonial governor and insisted upon the importance of their position as freeholders, independent farmers, heads of families, church members, and voters. This led to a certain aggressiveness in their attitude toward outsiders who might not appreciate their importance and worth, and an absence of formality from their manners that gave foreign observers an impression of rampant democracy. An English official reported of inhabitants of rural Connecticut, "They seem to be a good substantial Kind of Farmers, but there is no break in their Society; their Government, Religion, and Manners all tend to support an equality. Whoever brings in your Victuals sets down and chats with you." But in his own community, the yeoman farmer or artisan infused a thrifty, sober steadiness into the tempo of village life.[14]

Most residents of the towns owned some land. "Where there is one farm in the hands of a tenant," reported Governor Hutchinson, "there are fifty occupied by him that has the fee of it."[15] In like manner the number of indentured servants was small as compared with that of the middle and southern colonies. Few Negro slaves were to be found in rural New England where the prevalent system of husbandry was unsuited to forced labor. Actually members of "dependent classes" were few and, as the author of *American Husbandry* found, better off than their brothers anywhere else in the world, because their labor brought high wages and cheap land was generally available to them.[16]

So much for the Yankee men. What of their mothers, wives, and daughters, the forgotten women who, after all, made up one

half of the town community? Even in the eighteenth century no
New England male would have denied the usefulness of woman.
He was strong to wive, because he needed a woman to keep his
house, produce his labor force, cook his meals, spin his flax, dip
his candles, care for his garden, minister to him when he was sick,
and perhaps even aid him in the fields. A good wife would husband his small surplus, increase his standing in the community,
and not rarely provide him with companionship. Colonial women
deserve our tribute because of their remarkable energy and character. Despite excessive childbearing they performed feats the
mere thought of which would stagger their modern sisters. The examples of three good wives of Windham, Connecticut, will suffice.
Hannah Bradford cared for the town sick for many years and is reported to have taught the first male doctor of the community much
of his medical lore. Mary Howard kept her own house and that of
John Cales, whose estate she managed most skillfully. Lucy Reyn-
olds could kill a bear, heft a barrel of salted meat, or, if challenged
"throw the strongest man in Windham." In industry and frugality
they equaled the New England man, whose protection they did
not necessarily require. Mrs. Hannah Mackerwethy of Dedham,
Massachusetts, started life with a legacy of £20 from her father and
bequeathed to grateful heirs at her death, in 1771, £1,200 "which
she had earned with her own hands by spinning." She had at the
same time made herself a power in the community, giving generously to the support of church and minister.[17]

President Eliot of Harvard once remarked that the New England woman had not only to put up with her work and her children, but she had also to put up with the New England man.
Legally her husband was her lord and master. He controlled all
property she possessed at the time of her marriage and whatever
she might acquire later, and all family decisions were made by
him. We must make careful distinction between colonial woman's
usefulness to her man and the amount of social recognition he accorded her. However, rebellion against his smug tyranny was not
unusual, and advertisements for runaway wives appear in the
newspapers with startling frequency. "My wife Annabelle Holman," announced John Holman of Bridgewater, Massachusetts, in
1745, "absolutely refuseth to cohabit with me at my house in said
Bridgewater, but keeps still at Newport, where she is continually

running me into debt and exposing me to many lawsuits . . ." In Annabelle's case we note a quite understandable preference for the bright lights of the "Metropolis of Southern New England" to household drudgery among the local yokels of Bridgewater![18]

It is extremely doubtful, however, that the average New England housewife regretted her lot. In the end, life was managed pretty much as she wanted it, for her role in the scheme of things was too vital for her to be long disregarded. Throughout the ages her sex had learned that where the frontal attack on the male often failed, success could be attained by indirection. Frank men admitted this. When Jonathan Brewster of New London wrote to induce John Winthrop to return to town because "Wee and the whole Towne and Church wants you," he simultaneously invoked the aid of Mistress Winthrop, for "you know weomen are very strong and powerful to act this way, and overcome the strongest and wisest men that ever were or are in the world, by perswations and swete alurements."[19]

In 1778 John Adams gave Major Langbourne of Virginia a recipe for making a New England town: town meetings, training days, town schools, and town churches. "The virtues and the talents of the people are there formed"; Adams went on to explain, "their temperance, patience, fortitude, prudence and justice, as well as their sagacity, knowledge, judgment, taste, skill, ingenuity and industry." There is no question that the New England character was given direction by its peculiar institutions, nor that its virtues (that they were as many as Adams listed I will not try to maintain) as also its defects stemmed directly from the same source. It will, however, be instructive to pass these agencies briefly in review in order to note their particular effect upon the developing philosophy of town life.[20]

"Town Meeting to Day makes a great nois and hubbub," Theodore Foster of Brookfield recorded on February 19, 1768.[21] This was not unusual, for inasmuch as the local franchise was much broader than that of the colony, nearly every man in the town had a voice and used it. At town meeting the Yankee could air his views and grievances and explain why the town should not spend money for this or that project. Fresh from a hard struggle to make both ends meet, the freeholding farmer was a conservative and habitually suspicious of innovation. "He watched public expenditure with a cold, saving eye, and in town meeting could be safely counted upon to raise his voice against anything which was likely to impose a burden on his acres."[22] In this respect he saw eye to eye with the large property owner and usually went along with him in voting "no" on appropriations; when it came to voting taxes, he was downright tight. If money must be spent, he wanted himself to have the voting of it.

Political caldrons boiled only a local brew until the very eve of the Revolution. Minor town jobs, from hogreeve to fence-viewer, were all filled by farmers, who in a real sense rendered yeoman service. Where town matters were involved, politics aroused the greatest interest. The people of a New England town managed its affairs and wanted no interference from without. Beyond voting instructions for the town's representatives in the colony's assembly on such pertinent questions as an issue of paper currency or a new road from the metropolis, the townsmen were generally apathetic on provincial or imperial questions, an attitude the imperial authorities never quite understood.

Yet this tiny forum graduated many a hard-headed political thinker, in whom, despite ingrained village prejudices and ample misinformation, analytical acumen and capacity to lay bare the fundamental principles of government combined to a degree surprising to the present age. Such a Yankee was William Manning, who grew to manhood in the Old Manse at Billerica and nourished his deep-seated distrust of lawyers, aristocrats, and government on newspapers, the oratory of Otis and Sam Adams, and talk. "Teased in his mined . . . for many years" with what seemed a menacing disregard of the great principles that had taken him to Concord with the minutemen in 1775, he painfully composed in 1796 a shrewd, vigorous, original and salty, if unlettered, tract for democracy, *The Key of Libberty: Shewing the Causes Why a Free Government has always failed and a Remidy Against it*. In this unique and precious document we glimpse the profound truth of the assertion that "with the sole exception of England of the Commonwealth, no community in modern history has been so fecund in political thought, as America of the revolutionary generation."[23]

"They are all Politicians," said James Hulton, "and all Scripture learnt." This juxtaposition ceases to surprise when we remember that the New England meetinghouse served both as town hall and house of worship. It was never called a church, for as Richard Mather had said, "there is no ground from Scripture to apply such a trope as church to a house for public assembly."[24]

Here, then, was an outward and visible sign of the curious mixture of the ideal and the practical in the Yankee character. This being so, it was the church as a social rather than as a religious institution that was so important in shaping the lives of townsmen. Theological emphasis may have been strong at first, but after the embers of the Great Awakening died down, about 1750, it is probable that the average New Englander would rather have wet his whistle with a gill of rum than sweeten his mouth with a morsel of Calvin.

Church attendance on the Sabbath by no means turned out to be the chore that some historians of our godless age imagine it to have been. In truth a good time was had by all; even a youth like Thomas Mentor of Ipswich claims the attention of posterity because he was reprimanded for "taking of the maids by the aprons as they came into the meeting house . . . putting his hand in their bosoms," snatching away their posies, and laughing and whispering with "the very little boys" all during the time of worship "most of the Sabbaths of this year." For the elders, now that singing by note had been introduced, the rendering of hymns proved a pleasure, although the parson still had occasionally to chide his congregation for ignorance and heedlessness "in sliding from one tune to another while singing or singing the same line in different tunes."[25] New England ministers had to be good to hold their jobs. The General Court of Connecticut stated inclusively:[26]

"That by an able and orthodox minister . . . they understand a person competently well-skilled in arts and languages, well-studied and well-principled in divinity, approving himself by his exercises in preaching the gospel to the judgment of those that are approved pastors . . . to be a person capable of divining the word of truth aright, to convince gainsayers, and that his conversation is such that he is a person called and qualified according to gospel rule, to be a pastor of a church . . ."

Those who did not suit were liable to acid remarks from village critics, like the brother who declared, "I would rather hear my dog bark than hear Mr. Billings preach."[27] As secularism filtered slowly but pervasively into the puritan way of life in the mid-eighteenth century, the clergy shifted from theological and doctrinal sermons to dissertations on social and political subjects. Many an unlettered Yankee absorbed by ear from the discourses of a learned minister much knowledge, especially in the field of political theory, and, by the time the Revolution threatened, had become thoroughly familiar with the jargon of the pamphleteers. One loyalist blamed much of the rebel spirit of New England upon members of "Sam Adams' black regiment" who allowed their pulpits to "Foam with Politicks."[28]

In addition to providing townspeople with an opportunity to assemble, exchange amenities, and listen to sermons twice a week, the church was a powerful instrument for social discipline, and played an important part in the regulation of morals and conduct. Everyone, whether religiously-bent or not, felt the influence of this institution in the life of the community. In the social rather than the theological sphere lies the lasting significance of the New England churches.

Discussion of the contribution of militia training and the school to New England town life may be passed over rapidly. There was a certain disciplinary value to the former; it developed a local *esprit de corps* and gave a rudimentary training in leadership and cooperative effort to all concerned. Every man from sixteen to sixty had to be a citizen soldier. Officers' interests and ambitions centered often on election and promotion and on the increasing prominence offered by good regimental connections rather than in provision for the common defense. The colonial records of New England are literally stippled with military titles. Dr. Alexander Hamilton and a friend, being in Salem in 1744, one evening dropped in at the Ship Tavern, "where we drank punch and smoaked tobacco with several colonels; for colonels, captains, and majors are so plenty here that they are to be met with in all companies. And yet," the physician shrewdly observed, "methinks they look no more like soldiers than they look like divines; but they are gentlemen of that place, and that is sufficient."[29]

With the quieting of the frontiers after the last French War, the militia became more than ever a local levy and another example of the New England town's exclusiveness and self-sufficiency. The New England farmer might keep his musket ready for the protection of his home, but he resented being summoned to the defense of a neighboring community, and, since soldiers cost money, preferred to stand guard in his own town unassisted. Though the gentry may have prayed to be blessed with ample rum and military titles, they had no hankering after military glory, and recruiting officers continually found poor pickings among local militia. Eight years of fighting for independence hardly altered this parochialism. Muster day provided occasion for rough sociability eight times a year but entirely lacked the pomp and formality that characterized parades of uniformed cadets or fashionable troops of horse in more urbane communities.[30]

Of the school, it may be pointed out that poverty and the need for the labor of every pair of hands throughout the period prevented a realization of the New England ideal of a school for every forty families and a Latin school for every hundred. Yet by 1750 most wealthy communities maintained a grammar school for at least part of the year, and in virtually all of them ambitious lads could read with the minister for entrance into college. At Woodstock in Connecticut, where drain of money and manpower during the French and Indian wars defeated all efforts to maintain a grammar school, the Reverend Ezra Stiles prepared several young men for Yale. Indeed there was throughout New England a great respect for learning as one of the better things of life and a means of elevating a man to the professional and governing classes. Many a Yankee farmer, at great cost to himself, sacrificed the labor of a promising son to allow him to acquire a Harvard or Yale degree. It is probable that a majority of townsmen could read and write; at any rate they respected those who did and, where they were able to raise the funds at all, were quite willing to devote them to a school.[31]

To these four institutions singled out by John Adams as the principal formative agencies of the New England character, I would add for these years at least two more—the tavern and the highway system.

In any chronicle of the rise of republican government in New England, the town meeting, church, and tavern form a sort of trinity. And the most democratic of these was the tavern. "Here," remarked John Adams caustically, "vicious habits, bastards and legislators, are frequently begotten." Its joys were open to members of all classes, some of whom occasionally forgot their places in the social scale under the influence of the cup that cheers. To this center of village society came post and stage riders, and all "foreigners" having goods to sell or news to impart. At the inn much of the village business was transacted, and here in the long evenings the political engine was worked to the limits of its capacity. Tavern clubs became tiny forums for the discussion of all topics from local elections and the state of crops, through current deistic or literary controversy, to the latest rumor that Mary Parker and Obadiah Bartlett must shortly stand before the congregation and admit having indulged in carnal wickedness. Hours spent around tables with mugs of rum, mulled wine, or potent flip gave Yankee townsmen an education in human nature that the combined efforts of church, school, and even the colleges at New Haven and Cambridge could not impart. That New England inns became so famous even President Dwight felt impelled to sing their praises was in large measure due to the superior human qualities of the innkeepers. John Adams knew them well from his experiences while riding the judicial circuit and insisted that many of them were the "grandest people alive." The landlady at Treadwell's Inn, Ipswich, was a great-grandaughter of Governor Endicott, and had "all the great notions of high family that you find in Winslows, Hutchinsons, Quincys, Saltonstalls, Chandlers, Leonards, Otises," and Winthrops, while her husband was "as happy, and as big, as proud, as conceited as any nobleman in England . . . ." No man, be he even the parson or the squire, cut a finer figure than mine host of the village tavern, and few knew more about American conditions or wielded more influence in town councils.[32]

After listening to tap-room poetry for a lifetime, John Daggett of Attleborough, who kept a tavern on the Boston-Providence road, composed his own epitaph, "addressed to a Traveller," which was printed in the *Newport Mercury*, June 27, 1768:

> If e'er good Punch to thee was dear,
> Drop on John Daggett's Grave a Tear:
> Who then alive, so well did tend
> The Rich, the Poor, the Foe, and Friends,
> At ev'ry Knock, and ev'ry Call,
> I'm coming, Sir, he cry'd to All:

At length Death Knock'd! poor Daggett cry'd,
I'm coming, Sir! And so he dy'd.

Between 1740 and 1776 a network of passable highways slowly spread over all New England.[33] Poor, but adequate for their day, they played a paramount role in the breakdown of rural isolation.[34] After 1750 stage routes operated on regular schedules between important points, and post riders carried the mails and newspapers to rural areas with considerable frequency.[35] In addition peddlers with their mysterious packs came periodically and increasingly over the roads to compete with the local general storekeeper. By this period, too, rural customers could purchase by mail order from Boston nearly any article desired. Samuel Hardcastle, at the sign of the Three Nuns and Comb, in Cornhill Street, announced in his hardware advertisements that country customers "may be furnished by sending a letter as if present themselves." The same held true for wholesale orders of dry goods from the warehouse of Jonathan and John Amory. Even "Bride and Christening Cakes, ornamented in the Genteelest Manner at a Pistareen a Pound," might be procured from Thomas Selby's Pastry and Kitchen Cookshop for rustic festivities, though nothing is said about their condition upon delivery.[36]

More significant than the facilitation of trade by the improvement in transportation was the passage of culture over these roads, by means of which rural New Englanders began to share in the continuous process of civilization in transit from Europe to America. By 1766 Boston and Newport newspapers were regularly delivered by "News Riders" at taverns throughout most of rural New England; the *Massachusetts Gazette and Boston News-Letter* announced that "it circulates mostly in the Country, especially on the Great Western Road to Worcester, Springfield, Northampton, Hartford, New Haven, etc."[37] Boston maintained its importance as the greatest bookselling center in America by virtue of the expanding hinterland market for all kinds of reading matter from almanacs to novels, from the poems of Stephen Duck of Wiltshire to the writings of John Locke. As in earlier days peddlers served as book agents, but now booksellers also used the mail order system, whereby books and all sorts of printed matter reached remote destinations. Many a Yankee farmer purchased religious and political tracts, and even occasionally a work of fiction. There were practical works as well, like *The Cyder-Makers Instructor*, Culpepper's inevitable *Household Physician*, and *Look e'er you Leap: A History of Lewd Women*, in Chapter III of which we find "women considered in the threefold Capacity of Maid, Wife, and Widow.—With Directions how to choose a good Wife." The book was sober enough, but like our modern films the title promised more.[38] John Mein, one of Boston's leading booksellers, in 1765 launched a circulating library of several thousand volumes, from which his country subscribers might withdraw by mail two books at a time, and nine years later Henry Knox from his Book Store in Cornhill Street offered special prices to "those Gentlemen in the Country who are actuated with the most genuine Principles of Benevolence in their exertions to exterminate Ignorance and Darkness, by the noble Medium of Social Libraries."[39]

By 1770 the ideas coming over the high road to the New England town to hammer against its insularity and provincialism were slowly swinging the interior into the current of seaboard thought, and subtly preparing the minds of men for momentous events to come.

The people of New England and their institutions were gradually fused in the crucible of the New England environment into the folk known as Yankees. But what was their mental outlook, the philosophy that made for the town way of life? The New England character was, at one and the same time, extremely individualistic yet intensely social. This apparent paradox resolves itself when the New England way of life is subjected to analysis.

The Yankee's ideal was basically an agricultural ideal, and his outlook that of the farmer. Living close to the soil he knew instinctively the odors of field and forest, the songs of birds, the habits of animals. This intimate contact with nature continually reflected itself in earthy speech, even in the discourses of parsons, as when the Reverend Josiah Dwight told his listeners that "if unconverted men ever got to heaven, they would feel as uneasy as a shad up the crotch of a white oak."[40] The rural New Englander was never far from consciousness of the mutability of the seasons or the unchanging courses of the stars. Yet at the same time he did not, like other American colonists, live on an isolated farm but where he could enjoy the stimulating benefits of village society. His individuality was thus tempered with sociability from the start.

So rigorous was the climate and so unyielding the soil that the
New England farmer was kept eternally busy with the mere gain-
ing of a living. The easier husbandry of the South or of the unri-
valed limestone lands of Pennsylvania he knew not; he had to
make a virtue out of the realities of his hard life and continually
practice industry and economy. In matters of spending he could be
downright mean, but he did spend when necessary; he spent well
and got the best. Lovely churches, substantial and well-propor-
tioned houses, and taxes for education stand witness for him. New
England thrift was zealously contemplated, but it was never inef-
ficient. And no one who genuinely deserved charity ever met with
denial from inhabitants or town. Yankees were of one mind about
shiftlessness and waste; they were detestable if not actually dangerous, as threatening the very basis of existence. So too their
handmaidens, ostentation and display.

Nothing is more misleading than the assertions of some superfi-
cial historians that town life was dull and unedifying. On the contrary, existence in the villages was at times decidedly hilarious,
vulgar, often noticeably intemperate, and usually enjoyable. A free
and generous hospitality prevailed, good cheer abounded, and
rural fun and frolic were often indulged. Being human, the Yankee
was a sociable animal, and his far-famed humor was never intended or suited for soliloquies. The critical curiosity of outsiders
coming into town often met with a cool welcome, and their reports
like those of their fraternity in all ages, may be somewhat
discounted.[41] New Englanders had plenty of fun of the spontane-
ous sort common to all rural societies, and their capacity for self-
entertainment was indeed remarkable.

Town living fostered development of what we call the "town
character," since the slower tempo of life in a compact rural setting
bred a great variety of human nature. Eccentricity was not only
tolerated by the community, it was often actually encouraged because it made people more interesting. It was a luxury a person
certain of his social status could afford to indulge. Besides a man's
peculiar habits and tastes were his own business. No check to the
development of individuality was needed or applied so long as it
did not become anti-social or interfere with the community ideal.

An outstanding attribute of rural New England in the mid-
eighteenth century was its humanity, nowhere more evident than
in the attitude toward morality and sin. True people did mind

one another's business, but in what society in any age do they not? Yet it was more than mere friendly or idle curiosity that led to snooping and prying; for the good of the community the social ideal must be preserved. The church still held before the people a stern moral code, but its puritanism had noticeably mellowed. At the tavern or in village homes most honest folk silently agreed that the flesh was weak; they and their ministers wisely realized that a frigid asceticism could find no more place in New England than in any society close to the soil, but they nevertheless set themselves a high mark to shoot at. The majority approved a strict morality, yet inclined to leniency with sinners who fell by the wayside. Sex was a private matter so long as it did not cost the community anything. Couples guilty of sexual lapses were required to admit their sin on the Sabbath in the meetinghouse before a full congregation, but this could hardly have proved a harsh penalty, since church members were all neighbors, and in a small community such situations were probably pretty general knowledge before ever they saw the light of public confession. "These confessions were very frequent," says an unimpeachable authority, and the seven months' baby became a Yankee by-word.[42] That any stigma permanently attached to what all regarded as a temporary youthful lapse is not borne out by the future attitudes of the town. Adolescent wild oats were easily forgiven and quickly forgotten when the young man and young woman indicated their willingness to settle down and assume their share of community responsibility. The town was far more likely to deal harshly with those whose loose utterances might threaten the social order. When Jeremiah Ripley of Windham publicly asserted on election day at the Court House "That the Honorable Governor was a fool, and his friend and counsellor, Roger Walcott, a knave, and that we will turn out the knave and kick out the fool," his penalty was the loss of his franchise, the posting of £100 bond for good behavior, and payment of costs.[43]

An orderly place was the average New England town. When Bryan Sheehan of Marblehead was found guilty in Essex County Court in 1771 of rape, the press announced him to be "the first Person, as far as we can learn, that has been convicted of Felony in this large county, since the memorable year [of the witchcraft], 1692."[44] Yankees made honest and well-behaved citizens, not only because the church taught that such demeanor was enjoined by

Holy Writ, but because the town as a social unit functioned better that way. In their transactions with one another the historian can find little to arraign. But when they left their own locale the lid came off, and the Yankee earned the reputation for sharp practice pinned on him by the people of colonies to the southward. Of course you dealt honestly with your neighbors, but if you could put it over on the city folk of Boston you were rather a smart fellow, and did no real harm. In 1758 a farmer sold a turkey in Faneuil Hall market inside of which the purchaser found a brick weighing fifteen ounces, and another "Countryman was detected in the Market, for selling a Quarter of Meat, which was found to be stuff'd under the Kidney with some empty Guts, etc., to set it off the better. . . ." In this case the sin consisted in being caught, and when fined twenty shillings, the culprit "with Great Humility" promised reformation. Amos Brown of Stowe made a practice of concealing putrid butter beneath a covering of good for the benefit of his city customers till a fine of £6 and the posting of a large bond wiped out his profits.[45] There is historical precedent in plenty for the wooden nutmeg racket and the candid sign on display in a New Hampshire town: "Antiques, made and repaired."

In common with the rest of New England, rural life shared in the growing secularism of the eighteenth century. Slowly but perceptibly, the old socio-religious emphasis gave ground before the more mundane commercial point of view. Even the clergy were swayed by the new winds of trade. With salaries none too adequate at any time for the social plane on which they lived, some of them devoted large portions of their time to secular concerns, such as speculation in town lands and other business enterprises. Frequently parishioners were shocked by such worldliness, as at Gorham, Maine, in 1757 when the congregation complained to the town proprietors that the Reverend Solomon Lombard "has taken upon him so much business which does not concern the ministry, which gives us grounds to think him more for the fleece than he is for the flock."[46]

From the first there had been in the New England character a conflict, or at least a dualism, of practical and ideal. To this the Calvinistic system, which tended to find in worldly success an indication of spiritual salvation, contributed a moral sanction. As soldier, colonist, or farmer, the New England Puritan believed he should trust God but keep his powder dry. With the material leavening of the eighteenth century, his philosophical outlook developed a growing dichotomy as puritan idealism had increasingly to contest for supremacy with Yankee practicality. In a letter accepting the pulpit of the Braintree church, the Reverend John Hancock expressed his deep sense of the "seriousness, solemnity and affection" of the occasion and his quite proper feeling of having been called by the Lord to the new charge. Then he suddenly added, "I would just take leave to recommend to your consideration the article of wood, which I understand is, or is likely to be pretty dear and scarce in this place." [47] The very realistic approach to questions of business honesty and village morality already mentioned reveals other aspects of this conflict between a noble ideal and the practical necessity for getting on in the world. We have no just cause to brand the Yankee a hypocrite for this attitude; here again his struggle for survival in an ungracious country, coupled with his broad knowledge of human nature, had taught him that God is far more likely to help those who help themselves. New Englanders took the position that religion (and this includes idealism) was a joint stock partnership between themselves and the Deity. They were willing to take God in on what they were doing, but first of all they wanted to save themselves.

It was this attitude of practical idealism that made the needs of the town more imperative than those of the individual. The Yankee was a political and social being mindful of his responsibilities to his fellow townsmen. He shrewdly recognized that the town as a body was wiser than he, and that in the long run what was best for the community was probably also best for him. The town allowed the individual to develop any number of personal idiosyncrasies so long as these did not threaten the structure of the social order, but nothing was suffered to attack family or church, business or social institutions. In the eighteenth century a broader view of what constituted good citizenship succeeded the narrow "rule of the Saints," but the fundamental ideal of the town as a social entity existing to foster and promote a way of life prevailed as long as the town possessed any vitality at all. "Keep up your place," said the town, "support church, schools and town government, if not actively, at least by declining to undermine them. Keep your children off the town, mind your own business, and don't annoy your neighbors. Beyond this, you may go your own way." The Yankee, under such a dispensation, appears as a rugged individualist only within certain very definite limits.

Springing from its very nature, its ideal of self-help and self-sufficiency, came the greatest weakness of the New England town. It bred in its citizens an intense and at times insufferable provincialism. Yankees ran their towns efficiently, and were attached to their way of life, which they rationalized themselves into believing the only way possible and right. It followed, therefore, that they had nothing but contempt for those benighted souls to the south and west who did not order things as they did in New England. They could bow to the social ideal of the town, but not to that of the nation. Their outlook was confined to a narrow area six miles square. It is profoundly significant that prior to the revolutionary period few provincial leaders rose from rural towns. Such leadership as the country party enjoyed came from Boston bosses like the two Elisha Cookes and the elder Samuel Adams. Only with the opening up of roads and the extension of the postal system did this excessive localism begin slowly to broaden, and it may plausibly be argued that the New England town achieved its greatest influence, both as a producer of leaders and as a mother of new communities in New York and the West, at a time when it had already begun to lose its perfect adjustment as a social and economic unit.

But in its day the New England town, even with this serious limitation, was a success. It succeeded because a group of like-minded men and women, with a common religious faith, a common political outlook, and a common agricultural economy, were willing to subordinate themselves as individuals to the social ideal of a well-regulated community.

VIII

Violence and Virtue
In Virginia, 1766;
or,
The Importance of the Trivial*

I

THE KILLING

Mr. Robert Routledge, a merchant who lived south of the Appomattox River in Prince Edward County in his Britannic Majesty's Royal Colony of Virginia came from Cumcrook in the parish of Stapleton, county Cumberland, not far from the Scottish border. In the eighteenth century the Cumbrians had rather more in common with the Scots than with their fellow Englishmen to the southward. Cumcrook lay within ten miles of the Riddings where the forces of Bonnie Prince Charlie crossed the river Esk into England. We cannot say whether Robert Routledge joined the rebel army or whether he was an Anglican or Presbyterian. We do know that he came from a prosperous yeoman family—his mother was the daughter of a gentleman—and though a bachelor, he left a bastard son when he migrated from "his country" to America. Perhaps, like his brother Thomas, who amassed a fortune in the service of the Honorable East India Company, he merely went in search of opportunities. He arrived in Virginia about the year 1746.[1]

*This essay is reprinted by permission of the Director from the *Proceedings of the Massachusetts Historical Society*, Volume LXXVI (1964), 3–29.

In partnership with John Pleasants, Routledge prospered in trade and acquired some 1,272 acres of land in Prince Edward County; possession of the land made him a gentleman of Virginia. He won esteem as "a worthy blunt man, of strict honesty and sincerity, a man incapable of fraud or hypocrisy" and at the same time was very popular. By 1766 he enjoyed the reputation of being one of the substantial resident merchants of the Old Dominion, one who could count among his friends and cronies such Southside leaders as the Carringtons, Hartswells, and Tabbs, not to mention such Peninsula worthies as Colonel John Chiswell of Williamsburg.[2]

"Pretty early in the morning" on Tuesday June 3, 1766, Robert Routledge and some gentlemen of his acquaintance met in Benjamin Mosby's tavern at Cumberland Court House, and there they spent the greatest part of the day. Toward evening (as Virginians called the late afternoon) Colonel John Chiswell[3] dropped in, and after some time he began to talk "in an important manner," larding his discourse with "illiberal oaths." "With less politeness perhaps than was due to a man of Colonel Chiswell's figure," Routledge, in an artless way, expressed his disapproval. Thereupon the Colonel became abusive, calling the merchant "a fugitive rebel, a villain who came to Virginia to cheat and defraud men of their property, and a Presbyterian fellow." Thus provoked, the Cumbrian, who by this time was thoroughly drunk, threw some wine from his glass at Chiswell's face—"some small part of which did touch him" according to a newspaper account. In the code of the eighteenth century "this was an indignity which perhaps men of honor ought to resent from any one, unless from an aggravated and abused friend, or a man intoxicated with liquor." These two gentlemen had been "intimate" friends, and John Chiswell was "perfectly sober." Nevertheless he attempted to throw a bowl of bumbo at Routledge but was prevented from so doing by one of the company; he next endeavored to throw a candlestick at him but was again thwarted and then tried to strike Routledge with a pair of fire tongs, but again he was prevented from carrying out his intentions. Thoroughly enraged by this time, Chiswell ordered his servant to fetch his sword from a nearby house and threatened to kill the lad if he failed to comply.[4]

In an unusually complete and accurate narrative of the Routledge-Chiswell affair in Purdie & Dixon's *Virginia Gazette* of July 18, 1766, *"Dikephilos"* [5] provided colonial Virginians and posterity with the *first American murder diagram*, one which spelled out each fatal move and designated exactly the spot where the body was found. It will be convenient in re-creating the killing to use the letters given on this diagram.

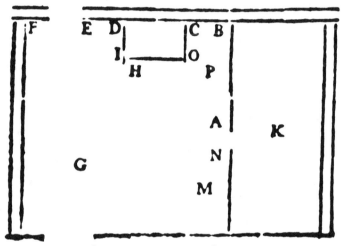

The First American Murder Diagram

Chiswell's boy returned to the tavern with the sword, which he delivered to his master, who was in the shed (K); whereupon the Colonel entered the public room and walked toward the chimney (M). Several of the company, perceiving the naked blade, came forward in the hope of persuading him to give it up, but Mr. Chiswell moved close to the wall (N) and "swore that he would run any man through the body who should dare to come near him or offer to take his sword. . . . Then he ordered in an imperious tone that Routledge should depart out of the room, as unworthy to appear in such company; and swore that if he did not immediately get out, he would kill him."

The intoxicated merchant, hiccoughing, insisted upon remaining in the tavern and said he held no grudge against his erstwhile friend. At this juncture, Mr. Joseph Carrington gingerly conducted his tipsy companion to the door at the back of the room (E-F) as

Chiswell, all the while fiercely abusing Routledge, moved along the wall (A to B). The distance from one principal to the other (B-E) was six feet, and a table (D-C) measuring three feet five inches stood between them against the wall. In order to retain his sword, Chiswell sought to get behind the table, and, while Carrington fumbled in his pocket for the key and Routledge stood by the back door (E), the Colonel concluded his vituperative remarks by repeating the words "Presbyterian fellow." At this, Routledge came forward to the table (I) muttering "fellow?" To which he added, "He thought himself as good a fellow as John Chiswell." He made no attempt to pursue Chiswell around the table—there were several gentlemen gathered about it who would have prevented him had he done so—but Joseph Carrington did actually grab him.

Immediately Chiswell, sober but highly overwrought, who had shifted from the wall to a position at P, now advanced to the corner of the table (O), "and with his sword, or hanger, stabbed him (I) through the heart across the table." Mr. Thompson Swann was standing close to the table (H), and the sword passed through his coat. Someone seized Chiswell from behind, but the Colonel avouched that it was too late: "He is dead, and I killed him." The Cumbrian, collapsing into the arms of his friend Carrington, died on the spot.[6]

Almost immediately, John Chiswell, "with the greatest calmness and deliberation," directed his boy to take his dripping sword and "clean it carefully and rub it over well with tallow, lest it should rust." Meanwhile Robert Routledge lay dead on the floor. The Williamsburg gentleman resumed his abusive outbursts, concluding with a theatrical declamation worthy of David Garrick: "He deserves his fate, damn him; I aimed at his heart and I have hit it." After this bombast, he "called for a bowl of toddy, and drank very freely," so that by the time the justice of the peace arrived "he was somewhat intoxicated." The justice examined all of the gentry present, among them Mrs. Joseph Carrington, and then ordered Chiswell off to the county jail. "He showed no sign of concern the next day, nor indeed for some time after," though he did inquire about Mr. Swann, fearing that he had wounded him.

After the coroner's inquisition, the prisoner was taken before a specially convened Examining Court, held July 9, in Cumberland Court House on Deep Creek and charged with "feloniously murdering Robert Routledge." Chiswell haughtily insisted that he "was in no wise guilty." Thereupon eight "reliable witnesses"—Littlebury Mosby, Thompson Swann, Jacob Mosby, James Mc-Daniell, Charles Scott, Joseph Carrington, Thomas Vines, and George Frazier—were sworn and examined, and the prisoner was heard in his own defense. John Fleming, president of the Court, informed Colonel John Chiswell that after careful deliberation the justices denied his demand to be released on bail and that it was their opinion that his case, which involved life and limb, "ought to be tried before the honorable the general court for the said supposed fact and thereupon he is remanded to gaol." In all, Chiswell spent seven days in the county jail, during which time the innkeeper, Benjamin Mosby, victualled him at a cost to Cumberland County of thirty-five shillings.[7]

In the custody of Jesse Thomas, an under-sheriff armed with the Examining Court's warrant, the prisoner was conveyed to Williamsburg. However, before Sheriff Thomas could deliver him to the keeper of the public prison on Nicholson Street near the capitol, his party was stopped, and Chiswell was taken from the officer and admitted to bail by three judges of the General Court "out of sessions." The judges, John Blair (President of the Council), Presley Thornton, and William Byrd III, basing their action upon legal advice given them by George Wythe and two other "eminent Lawyers," freed John Chiswell "without seeing the record of his examination in the County or examining any of the witnesses against him" who had been sworn to appear in Williamsburg for the trial. The bail was regular, however, for "four worthy Gentlemen" had entered into recognizance for Chiswell's appearance, £2,000 for him and £1,000 for each of themselves. There matters stood until July 18, when Purdie & Dixon's *Virginia Gazette* commenced a controversy that brought society in the Old Dominion to the verge of convulsion.[8]

II

THE VIRGINIA POLITICAL SCENE, 1766

In the ninety years following the collapse of Bacon's Rebellion, nothing had occurred to ruffle the surface calm of political life in the Old Dominion. The members of the tobacco gentry were firmly in control and were, in fact, steadily consolidating their power under the able leadership of John Robinson, treasurer of the province and speaker of the House of Burgesses. Here and there, however, one could discover eddies of discontent. The westward spread of tobacco culture and slavery across the Piedmont, almost to the foot of the Blue Ridge Mountains during the years after 1730 had greatly swelled the membership of the Virginia gentry and brought about a hitherto unrecognized shift in the colony's wealth and power. The principal offices and most of the patronage still lay in the hands of John Robinson and those Tidewater gentlemen who resided in the counties close enough to Williamsburg to permit them to attend Council meetings and other political conclaves, but the planter aristocrats who lived west of the Fall-line were not merely eager but determined to win a greater share of political power, a share commensurate with their greater wealth and growing importance. Similarly, certain Tidewater families, chief among them the Lees of Westmoreland, had grown restive and felt the loss of their former influence. But so well entrenched was the governing "oligarchy" and so popular was its leader that an act by the bungling ministers of King George III was needed to provide the insurgents of the Piedmont with the opportunity they needed.[9]

During the debates on the Stamp Act in the House of Burgesses, the existence of a group of "young, hot and giddy members" headed by John Fleming, George Johnston, Colonel Robert Munford, and Paul Carrington was revealed, and in a new member from Hanover, Patrick Henry, they uncovered an orator of the very first class. Their differences with the Robinson-Randolph group were not those of "radicals and conservatives," for they were all respected members of Virginia's governing class; rather the differences stemmed from the challenge inherent in the strong wording of Patrick Henry's celebrated resolutions of May 29, 1765, offered to "the establishment" by comparative parvenus for the first time in the eighteenth century. Won openly in the Planters' Parliament, this success by the men of the upper counties served notice that henceforth they must be reckoned with, and evidence that the Tidewater men were shaken by the victory is to be found in the rally staged a few days later to expunge Henry's fifth resolution from the House's *Journal*.[10]

If any attempt at accommodation was in progress between the two factions after the repeal of the Stamp Act, two closely related events occurring within a month of each other served only to exacerbate matters. The first of these was the sudden death of Speaker John Robinson from the "torments of the Stone" on May 10, 1766.

No man in Virginia was more highly respected or better liked than John Robinson. When therefore, shortly after his death, news leaked out that instead of burning the paper currency turned in for redemption, he had lent it out again to friends and that his estate was indebted to the colony for the huge sum of £109,335 9s. 2d., pointed questions and unresolved doubts as well as consternation spread rapidly among the citizenry. In November 1766, after much discussion and no little excitement, the Assembly separated the offices of speaker and treasurer, and for the first time in the Old Dominion's history, an act passed setting penalties for misapplication of public funds.[11]

These political repercussions consequent upon the shortages in Robinson's accounts have received extended attention from historians. What has not been noticed is the interlocking relationship of the Robinson affair and the Routledge-Chiswell murder case, together with the bearing of both upon the political fortunes of the Tidewater aristocracy.

III

COLONEL JOHN CHISWELL, GENT.

Late in the seventeenth century, Charles Chiswell emigrated from Scotland to Virginia. He must have formed good connections rather quickly, for by 1706 or even earlier he had become clerk of the General Court and thereby enjoyed intimate contacts with persons in power. In 1722 he began to submit a series of successful petitions for grants of large parcels of land in the counties of Henrico, Hanover, Spotsylvania, and Goochland. He was already possessed of 24,500 acres in 1735 when he received authorization for the survey of 30,000 more on Rockfish River in what was then Albemarle County. A year later the Scot and four other acquisitive gentlemen of Hanover were granted a 30,000-acre tract on the south side of the James in Amelia and Goochland counties. The ownership of so many acres provides a sufficient explanation for the appending of the word "Gent." to his name in the official records. At his death in April 1737, "Charles Chiswell, Gent." bequeathed to his son John a substantial estate consisting of a mansion house and plantation of 7,000 acres located on the South Anna River in Hanover and the other parcels of land already mentioned. He also left to his son a secure place in the crystalizing aristocracy of Virginia.[12]

Although he was only of the second generation of his family to reside in the Old Dominion, John Chiswell proceeded to move rapidly to the forefront of the provincial gentry. His landed wealth gave him a good start, and his marriage to Elizabeth, daughter of Colonel William Randolph of "Turkey Island," did the rest. In the very year of his father's death, John Chiswell was appointed to the Hanover Commission of the Peace; two years later we find his horse Edgecomb winning the race at the Williamsburg Fair, and he already held the rank of colonel in the militia. First elected to the House of Burgesses in 1742, he represented his county in that body for thirteen years, and then from 1756 to 1758 he sat as a burgess from the City of Williamsburg.[13]

Social success naturally attended and abetted political advancement. The four daughters of John and Elizabeth Randolph Chiswell all made brilliant "dynastic" marriages. Susanna, the eldest, became the third wife of Speaker (and Treasurer) John Robinson, a gentleman of her father's generation who was one of the richest planters in the colony. The next most influential burgess was a son of King Carter, Charles, of "Ludlow," Chairman of the Committee of the Whole House, who espoused the Chiswell's second daughter Elizabeth. Notwithstanding the implication of their father in the murder of Robert Routledge in 1766, the two younger girls also made excellent matches. In 1768 Mary, "a most accomplished young Lady," was wedded to Warner Lewis of Gloucester, and in 1770 Lucy became the wife of Colonel William Nelson of "Dorrill" in Caroline. Gaiety and open hospitality prevailed both at the rambling plantation house, "Scotchtown" in Hanover and in the Chiswell House on "the back street" (Francis) in Williamsburg, where one could always be sure of meeting the "Superior Sort." And doubtless the family occasionally gave balls in the Raleigh Tavern, which Colonel Chiswell and Dr. George Gilmer had purchased for £700 in 1752.[14]

In spite of his assured place in Virginia society, Colonel Chiswell frequently behaved with the uncertainty of the *arriviste*, or perhaps one can attribute his haughtiness to the lack of a proper balance between his sense of pride in his high station and the *noblesse oblige*, which ought to have accompanied it. He appears by nature to have had a very quick temper, especially when he imagined that he had suffered an affront to his dignity and rank. At any rate he was widely known as a testy and choleric man, as the following incident attests. Toward seven o'clock one "Moist Morning" in July 1750 at Chiswell's Quarter near Rockfish Gap, the Reverend Robert Rose was present in a party of gentlemen when they were "Surprizd by an impertinent Constable called Dinwiddie" who "served a Warrant on Mr. Chiswell" and "summoned Him to Mr. Reids about 10 miles back." Colonel Chiswell "refused to comply . . . and ordered the constable and complainants off the plantation for their Rude behaviour and provoking Him to speak disrespectfully of Mr. John Reid, a Magistrate [of Albemarle County], who behaved very well in the affair."[15]

Social and political prestige and huge grants of land did not, however, guarantee financial security in mid-eighteenth-century Virginia. John Chiswell was one gentleman who paid far more attention to agriculture and commerce than most of his contemporaries; he left politics largely to his all-powerful sons-in-law. But his concentration upon tobacco culture and the development of his tracts in the west did not spell prosperity for this planter any more than it did for most of the planting gentry. He was chronically short of cash and land-poor. A factor for a London mercantile house, Francis Jerdone, sold Chiswell "European Goods" worth £1,500 sterling, "packages included," in May 1754 for a "fine parcel" of tobacco grown at "Scotchtown." The deal pleased him so much that he wrote his principals he hoped for a larger consignment of the weed from the planter the next year. Within a few months, however, Colonel Chiswell was so sorely in need of ready money and loans were so hard to get in his province that he wrote the leading magnate of Pennsylvania asking to borrow £2,000. Here we glimpse, alas only momentarily, the striking fact that a certain amount of intercolonial financing had been going on for more than a decade or the Virginian would not have known where to turn. Chief Justice Allen of Philadelphia replied on October 2 that "at present we have among us ten Borrowers to one Lender: and very few of the latter chuse to lend any money without the Province; As they cannot readily at a Distance command a punctual payment of their Interest." In a neighboring colony the interest rate is 7 percent, he added, "Yet our People had rather have Six within the Province for the Reason above." The large sum Chiswell requests cannot be had for any security outside of Pennsylvania. "Indeed Money is not near so plenty as it was during the War [1744–48], and some few years after, at which time there was money Lenders, who now chuse to employ their Stocks in Trade."16

By 1756, according to a claim in law made many years after by Widow Mary Willing Byrd, John Chiswell was indebted to her husband, the third William Byrd, for "upwards of thirteen hundred pounds sterling," which were due chiefly on protested bills of exchange and bonds. Further, at this time, the Colonel owed his son-in-law "many thousands of pounds," and John Robinson prevailed upon his father-in-law to transfer to him, as trustee, lands on Rockfish River and in Hanover together with lots and warehouses in Williamsburg, the Raleigh Tavern, and 130 slaves of different ages and sexes at "Scotchtown." In September 1757, three tracts totaling 23,000 acres were advertised for sale in the *Virginia Gazette,* and on May 31, 1760, Chiswell sold certain of these properties, including "Scotchtown," to John Robinson.17

A release from this crushing burden of debt came suddenly for John Chiswell in 1759 with the discovery on New River in the County of Augusta (now Montgomery) of a rich lead deposit. "The Lead Mining Company" was projected for mining the ore by Governor Francis Fauquier, John Robinson, William Byrd III, and John Chiswell as equal partners. The political eminence of this quartet ensured the immediate survey of a thousand-acre tract surrounding the ore bank, but when they discovered that Dr. Thomas Walker and his Loyal Land Company had a prior claim to the land, they agreed to pay him £2,000 for a clear title. According to articles signed February 20, 1762, each of the other three partners agreed to reimburse Robinson £1,075 for "their respective shares" and for the sum that he had advanced to buy out Dr. Walker. Soon the Governor withdrew, and Chiswell, for the rest of his life, claimed Fauquier's share without ever adducing any proof of his right to it. This same year the partners sent Chiswell to England to consult with the mercantile house of Farrell & Jones in the city of Bristol. Assays of the ore proved satisfactory, and accordingly the planter contracted through these merchants for the services of William Herbert and other skilled workmen, as well as for material and utensils needed for the mine. Upon his return to Virginia, Colonel Chiswell resided at the diggings for the most part, receiving and paying money for the works and serving as its superintendent "without any satisfaction appearing in the accounts." On the other hand, John Robinson advanced nearly all of the funds invested in what had become known throughout the southern colonies as "Chiswell's Mine." At Robinson's death, the Lead Mining Company owed his estate £8,085 12s. 5d., more than was owed to any other single individual.[18]

Such was the sorry state of Colonel John Chiswell's personal affairs when his son-in-law suddenly died in May 1766; and such also was the state a month later when the father-in-law, in a fit of rage, murdered Robert Routledge at Cumberland Court House. He was hopelessly insolvent. Nevertheless, on June 23 he made what was to become his last will and testament for the disposal of what he contended was "a moiety" in the Lead Mining Company consisting of lands, slaves, carts, wagons, and utensils, after the payments of his just debts to William Byrd III and the estate of John Robinson. It is of compelling interest to learn that the "Honble William Byrd, and Presly Thornton, Esqrs" were designated executors of Chiswell's estate along with his surviving son-in-law, Charles Carter, Gent.," and Richard Randolph. Byrd and Thornton, councilors and judges of the General Court of Virginia, were sworn as executors on November 3, 1766, when Edmund Pendleton presented the will to the General Court for probate.[19]

IV

A FREE PRESS CREATES PUBLIC OPINION

From the establishing of the *Virginia Gazette* in 1736 until May 1766, the largest and most populous of the English colonies in North America was served by only one newspaper. The publisher was generally careful in his selection of news and tended to be amenable to official suggestions because he was also printer to the colony. As a result of the Stamp Act, Joseph Royle, the public printer, stopped the publication of his newspaper in December 1765; he died the following spring. When Governor Francis Fauquier notified the Board of Trade of this event on April 7, 1766, he remarked that "as the press was then thought to be complaisant to me, some of the hot Burgesses invited a printer from Maryland. Upon which the foreman to the late printer, who is also a candidate for the place, has taken up the newspaper again in order to make interest with the Burgesses."[20]

"Tired with an involuntary recess from business these four months past," Alexander Purdie & Company resumed publication of Virginia's newspaper (unstamped) on March 7, 1766, with the encouraging declaration that "the press shall likewise be as free as any Gentleman can wish or desire." If Purdie entertained any hope of forestalling the publication of a second paper, he was soon disillusioned, for on May 16, William Rind, formerly with *The Maryland Gazette*, brought out the first issue of a new *Virginia Gazette*, "Open to All Parties but influenced by None." Rind frankly made a bid for the patronage of the "hot Burgesses" in his leading paragraph: "To enter into a minute Detail of the Advantages of a well conducted NEWS PAPER [*and the first that has ever been Establish'd in this Province*]* would at any Time, be impertinent, but more especially at a Crisis, which makes a quick Circulation of Intelligence peculiarly interesting to all the AMERICAN COLONIES. . . . The Interests of RELIGION and LIBERTY, we shall ever think it our duty to support."[21]

Recognizing that, in addition to competition from Rind's new

---

* This sentiment, italicized by the present writer, is written in a contemporary hand on the sole surviving copy of this issue of Rind's newspaper, which is now in the New-York Historical Society.

sheet, his own lay under suspicion of being too "official," Purdie changed his company's name on June 20 to "Alexander Purdie & John Dixon," printers, and at the same time he completely reversed the former policy of the newspaper. It was certainly time to do so, for that very day "a Virginia Planter" living on the Potomac sent a letter to Jonas Green of Annapolis, printer, which reached a very wide audience when the *Public Ledger* ran it on September 4 in London and Samuel Hall printed it in his *Newport Mercury* for January 26, 1767. The communication opened with the statement that "the undue influence the press, in Virginia, has long laboured under, renders it unnecessary to make any apology for conveying to the public through the channel of your free paper [*The Maryland Gazette*], anything from thence that breathes the spirit of liberty." The race for the public printing was now on at Williamsburg, and since there were only five months before the Assembly would convene and make an award, the pace was swift indeed. Throughout the contest, Rind's and Purdie & Dixon's two *Virginia Gazettes* served up for the excited inhabitants of the Old Dominion what they had never before experienced, the sensations and sensationalism of a free press.[22]

The rival printers could not have chosen a more opportune moment to inform Virginians about current local affairs and to attempt to mold nascent public opinion. Members of the Assembly knew much about existing political problems as a matter of course, but the general public remained pretty much in ignorance, save on the level of unverified rumor, as in the case of the aftermath of the Stamp Act (the repeal of which Governor Fauquier proclaimed on June 9). George Mercer's charges that Richard Henry Lee had eagerly sought an appointment as Stamp Agent, the campaign of Robert Carter Nicholas to separate the offices of treasurer and speaker of the House, and the scandal of the Robinson accounts— all of these provided explosive issues, which the old official *Virginia Gazette* brought out into the open for the first time on June 20 as Purdie & Dixon contested desperately for popularity and the award of the public printing. The detonator was the earliest press account of the murder of Robert Routledge and the surprising release of Colonel Chiswell on bail, composed by an anonymous reporter. Was he one of the printers?

"I ask whether this act of the three Judges of the General Court be legal. If it is legal, I have nothing more to say. If it is not legal, then I ask whether the act of these Judges has not a tendency to overturn the laws and constitutions of the country, by their exercising an extrajudicial power and controuling the course of law in a case of the highest consequences to the safety of the subject? Whether the bail taken by these Judges in an extrajudicial manner can be liable on their recognizances, if Mr. Chiswell should not appear to take his trial? If they are not liable, whether it is not in fact a rescue under pretence of law, of a person charged with an atrocious crime?"[23]

So few concrete facts were offered by the anonymous writer and so eager was the public for news instead of rumor about this unsavory affair that "Dikephilos" sent a long account to Purdie & Dixon, which they published on July 18. "Some persons have ungenerously surmised that the influence of Mr. Chiswell's friends will prevent the truth from being published," so the Reverend Jonathan Boucher, the author, has determined to tell the whole truth and to name names which "will open the eyes of some well meaning men." This long and fascinating article and three more by the same "Dikephilos" (for the clergyman chose to remain anonymous) made three important contributions to the public's understanding of the discussion: (1) they supplied an unassailably accurate account of the murder and its aftermath; (2) they clarified some of the legal points raised by defenders of the judges' actions; and (3) above all, they dealt forthrightly and honestly as well as penetratingly with the dangerous implications for Virginia's welfare in a government by gentlemen for gentlemen. The facts of the case up to the admitting of Chiswell to bail have already been rehearsed; we may now proceed to examine cursorily the legal aspects of the bailment.[24]

In the Old Dominion nearly every planter was something of a lawyer, and not a few of the gentry had read law at the Inns of Court in London. Great, therefore, was the interest aroused by the Routledge-Chiswell case, which understandably became the principal topic of conversation in drawing rooms and taverns. Several gentlemen lawyers, employing pseudonyms or remaining anonymous, regaled the readers of both Williamsburg newspapers

during the summer months of 1766 with explanations of what the
law was and what it was not. An unsigned article published by Purdie & Dixon on July 11 described the "amazement" of the public
when it was informed through the press "that three Judges of the
General Court (out of sessions) have dared to do a most flagrant injury, both to Prince and people, by presuming to rescue a person,
charged with the murder of his fellow subject, from the custody of
a sheriff, who by an order of an examining court, was conveying
him to publick prison in Williamsburg; and that they had discharged the criminal from custody, under pretence of admitting
him to bail"! A pernicious precedent had been created by this
reserving of a "power of licensing homicides," to the judges of
the General Court. Had the judges and their advisers "deigned an
answer to the queries, made on this subject, in a former paper,
they would have had no trouble from me."[25]

This thrust drew blood from George Wythe, Esq., who had advised the judges in company with two other lawyers. He contended in a piece published August 1 that the Court of King's
Bench in England has the power to admit all defenders whatsoever
to bail, even for the crimes of high treason or murder. "The General Court are, equally with the King's Bench, judges of all high of-
fences; and in criminal cases their sentences are as decisive and
uncontroulable, for appeals and writs of errour lie not from them,
neither can bills of exception to their judgments be admitted."
This legal opinion was delivered without comment upon the pro-
priety of using it in the present instance, but, said Wythe, "I
believed it was right."[26]

When "Metriotes" supported George Wythe's position in a
missing issue of Rind's paper by citing Lord Mohun's case and Rex
vs. Dalton as precedents, "Dikephilos" answered him with page
references to contrary views from the law reports of Salkeld and
Strange. "Marcus Fabius" and "Marcus Curtius" joined in September to demolish "Metriotes." They pointed out that "The General Court hath no power but what it owes to acts of Assembly; it is
therefore in vain to ascribe to it the powers of the King's Bench,
unless it receive them from the acts of Assembly. . . . Your argu-
ments, in behalf of the late bailment, are altogether taken from the
powers of a Court, founded on uncertain authority, in another
Country. . . . The laws of England are, I presume, in force in this

country; but there may possibly be some difference between the King's peculiar Court and the Supreme Court of this Country." They ended by charging "Metriotes," Wythe, et al., with subverting ideas of virtue and morality in the province under the guise of legal opinions.[27]

It is not necessary here to attempt to resolve finally the legal technicalities involved in the granting of bail. Evidently many lawyers believed the action of the judges to be extrajudicial as well as ill-advised. We can join with "Dikephilos" who said he could not censure George Wythe and his associates who gave advice "generally." Perhaps the differences of opinion among informed legal authorities arose from the situation in Virginia. Josiah Quincy, Jr., of Boston was an able lawyer who attended a session of the General Court at Williamsburg in 1773. He concluded that "The constitution of the Courts of Justice and Equity in this colony is amazingly defective, inconvenient and dangerous, not to say absurd and mischievous . . . It was a matter of speculation with me how such a constitution and form of judicial administration could be tolerable. I conversed with divers who seemed to have experienced no inconvenience and of course to apprehend no danger from this quarter; yet they readily gave in to my sentiments upon the subject, when I endeavoured to show the political defects and solecism of this constitution." It would be interesting to have Quincy's observations on the situation as it stood seven years earlier.[28]

During the summer months of 1766, Virginians of all ranks read more and more in their two newspapers about the actual workings of their government and society. To say that men were amazed at what they learned would be a considerable understatement. "Dikephilos" reported them to be extremely uneasy: "They say that one atrocious murder has already been cleared by means of great friends; and they are apprehensive that will not be the last opprobrious stain of the kind in our colony." He knew some citizens who feared that Colonel Chiswell would object to a "Cumberland Venire" and take care to have a menial or pliable tribe of by-standers planted ready to take their place" as the trial jury. He also learned that one or more witnesses to the murder had been tampered with. His immediate neighbors insisted that the law was being disregarded. "Patriots . . . are alarmed on this occasion; foreigners are alarmed; the middle and lower ranks of men, who are acquainted with the particulars, are extremely alarmed." Some of the deceased man's friends in Prince George County called for revenge for the killing of a worthy benefactor, but "Dikephilos" pleaded with them to remain quiet until the outcome of the November trial should be known. They agreed, but they also vowed "that if power exercise injustice and partiality, they can never permit the assassin, nor any of his abetters, to pass with impunity"—a threat that suggests lynching a generation before Judge Lynch.[29]

"Dikephilos" made a direct thrust at the abusers of privilege when he linked the murder with the Stamp Act. The latter was "justly detested," and many gentlemen, he stated, stood ready to give their lives to keep such oppression away and secure their liberties and property. "Now they apprehend that this partiality may be attended with still more dreadful consequences than even that detestable act of power could have been, because this must affect our lives, while that could only affect our estates." And who can blame them for being zealous for their security? Had Chiswell's crime been kept a secret, the cleric continued relentlessly, there might have been some excuse for "his friends who had no regard to the benefit of society, and were entirely void of patriotism, to endeavor to stretch points in order that they might save a man of Colonel Chiswell's figure." Even now many persons are apprehensive that Mr. Attorney's (Peyton Randolph's) "connexions with Col. Chiswell will occasion the prosecution for the King to be carried on in a different manner from what it ought to be."[30]

Several gentlemen of Virginia privately expressed the same opinions as those of "Dikephilos." One of these was the President of the College of William and Mary, the Reverend John Camm, whose comments to a gentlewoman in England ended with a condemnation of the three judges: they "have carried their power so high . . . which is like, as well it might, to put the whole country into a ferment." Robert Carter of Nomini Hall, member of the Council and therefore of the General Court, in describing this "melancholy incident," followed the account of "Dikephilos" very closely. He thought a conflict of testimony between John Blair and Jonathan Boucher "very alarming," but I shall neither applaud nor censure my bretheren's act, in the case just now. . . . I am not personally interested in the contest."[31]

Councilor Carter might not have been so unconcerned if he had read the newspaper of the previous day. In it was a letter addressed "To J[ohn] B[lair] Esquire" which declared that the admission of Chiswell to bail had not caused as much of an outcry as the manner of giving it. "All the public is concerned in is to examine how far itself is affected." Then the anonymous author pointed straight at the members of the Council: "In the present state of things, your fellow subjects in Virginia live only at the discretion of your Sublime Board, *a Board having an unreasonable power by law already*, should at least be prevented from usurping one, subversive both of law and reason." The fact that the judges took the advice of three eminent lawyers, all friends of the accused "selected by Colonel Chiswell's friends," has no weight with this writer, because the triumvirate assumed "a Regal power." Attorney General Randolph was not even present for the depositions of the so-called witnesses. When the relation between the county courts and the general courts, established by the Assembly, is altered in such a way, that "is to effect a revolution; we become another people." Thus was government in Virginia publicly called into question![32]

Some aroused Virginians began to voice their doubts about the social usefulness of an aristocratic ruling class. From Prince George County, Richard Hartswell communicated to Purdie & Dixon his agitation over two matters: John Robinson's breach of the public trust, and the fact that the Speaker's father-in-law was accused of murder. He had met both of these gentlemen in 1738. On a number of occasions he had stayed at "Scotchtown" for several days at a time, and then, at least, the Colonel was far from "liberal of oaths." However, if what Hartswell reads in the newspapers be true, then he firmly believes in carrying out the course prescribed by law. "Men in authority, especially the rich, have it more in their power to be of service to the public than any others," but they cannot justifiably be deemed "great men" if they fail to act properly and permit themselves "to connive at vice." He hoped his letter would reach the printing house, because less than two years before he had sent five letters to Royle's paper expressing his concern over similar problems and doubted if more than two had ever been received; the others were broken open and suppressed. "This scandalous practice . . . is too common to be looked upon as strange."[33]

Back in May before the death of the Speaker, Governor
Fauquier had written home that "Every Thing is become a Matter
of heat and Party Faction; every thing is contested; a Spirit of Discontent and Cavill runs through the Colony." At this time these
tensions were confined chiefly to the colony's rulers, but by the
end of the summer, the two newspapers spread them far and wide
to all levels of society. The first gentlemen were unaccustomed to
and ill-prepared for playing out their political and social roles in
such an open forum as that now provided. Amazement and fright
seized many of them when an open press published the naked
truth—frequently with acid comments. The historian who is familiar with the extremism of colonial newspaper controversies over
religion, as in New York and New England, must conclude that
this one in Virginia was, on the whole, restrained and focused
upon the essential issues.[34]

This novelty of printing the news that surprised and worried
the gentry fascinated and inspirited the populace. In August both
newspapers brought out a supplement containing "A Prophecy
from the East," which, in mock Biblical fashion, attempted to sum
up what had taken place in sixteen verses, of which a sample will
suffice:

3. Party shall menace Party, and Dunce shall enflame
   Dunce, and the Gazettes of *Purdie* and of *Rind* shall contain Wonders.

4. And it shall come to pass that the Principles of Judging
   shall be perverted; Mens Understanding shall be
   darkened: For this shall be the Aera of Delusion.

5. And then Men, who cannot write for themselves, shall
   write for the Public: The Great Men, even the Men of
   Fortune, shall write controversies.

6. And they shall call themselves Lovers of Truth, and
   Lovers of Justice: and Much Paper be wasted, and
   Words shall lose their Meaning.[35]

Such a state of public excitement was bound to provoke incidents,
two of which soon brought all issues to an end.

Colonel John Chiswell had departed for his mine shortly after
being freed on bail, but, true to his obligation, he returned to his

Williamsburg house on September 11 to prepare for his trial. There were many Virginians who believed that he was ready to admit his guilt but was reluctant to do so, exceedingly reluctant, as an aristocrat, to be hanged. Probably nowhere in the Old Dominion could an executioner with a suitable axe, let alone a silken cord, be found; and the accused man was too insecure a gentleman to submit to a method of punishment which was accorded to felons, felon though he was. Having read the newspapers carefully, both he and his friends were doubtless aware of ominous threats about what might occur when the trial began, which would be held without the benefit of any kind of police protection. They had good reason to fear mob action if they tampered with the jury or if the populace should suspect that the trial was in any way fixed. Here indeed the entire colony confronted a dangerous situation, one fraught with all kinds of unanticipated social and political implications.[36]

All ended with a grim and ironic finality, however. The General Court was to try the case on November 16–17, when the new term for the Assembly should commence, but on October 17, Purdie & Dixon's *Virginia Gazette* carried the following brief notice: "On Wednesday last, at 11 o'clock in the afternoon, died at his house in this city, Col. John Chiswell, after a short illness. The cause of his death, by the judgment of the physicians, upon oath, were [*sic*] nervous fits, owing to a constant uneasiness of mind."[37]

The sworn statement of the physicians about the "nervous fits" notwithstanding, the rumor went abroad that Colonel Chiswell, who, indeed, must have suffered constant "uneasiness of mind" as he reflected upon his guilt and the public's insistence that justice be done, had committed suicide. It is entirely plausible that, given his choleric nature, he may have died of hypertension or even of a coronary thrombosis or of some other ailment brought on by the great fear that possessed him. It is entirely plausible, too, that he either hanged or poisoned himself rather than stand trial, be judged guilty, and inevitably be hanged as a felon. His friends had already done too much for him; they could do no more in the face of an angry public. To this very day the historian can only say with Jonathan Boucher that John Chiswell, in an "extraordinary manner was found dead, it was never known how."[38]

If the frequenters of "Public Times" at Williamsburg were thus

denied on November 16 the spectacle of a gentleman killer on trial
for his life, they were provided on the following day with a court
case involving two of the same actors, a case, which in the long
perspective of history, has proved to be of far greater importance.
William Byrd III and John Wayles preferred a bill of indictment in
the General Court against Colonel Robert Bolling, Jr., of Buck-
ingham, one of the Piedmont leaders, for inserting in Purdie &
Dixon's sheet on July 11, a piece libeling Judges Blair, Byrd, and
Thornton for freeing Colonel Chiswell on bail. At the same time,
Wayles preferred another bill against William Rind, Alexander
Purdie, and John Dixon for printing in their newspapers the pre-
vious Friday "a piece signed R. M. and addressed to Mr. John
Wayles." A series of libel suits seemed imminent, "but the Grand
Jury this Day returned the said Indictments, NOT TRUE BILLS."
Whether he copied it from Rind's *Virginia Gazette* or wrote it him-
self, Jonas Green appended to his story the names of the members
of this blue-ribbon jury: "This Righteous JURY was composed of the
following Gentlemen, Mann Page, Foreman, John Page, Lewis
Burwell, senior, David Jameson, William Stevenson, Jerman
Baker, James Hubbard, Thomas Knox, Haldenby Dixon, Thomas
Newton, junr., Dolphis Drew, Robert Smith, Robert Shields, Wil-
liam Digges, junr., Esqs., GOOD MEN and TRUE, FRIENDS TO
LIBERTY."[39]

The plaintiffs did not stand alone, however, but represented a
substantial segment of Virginia's aristocratic opinion, as an extreme
statement signed "R. R." reveals:
"I am told that subordination is proper and necessary in all societies, and I am sure nothing entitles a man to superiority so much
as family and fortune. Therefore as some men who are inferior in
fortunes have undertaken to judge and blame their superiors in
this respect, I think that every *great man* must be affected with indignation, and endeavour to suppress this growing evil. I would
therefore recommend and insist that indictments should be speedily presented against five Writers, fifteen Evidences, three
Printers, two Sheriffs, and twelve Veniremen, that they may suffer
for their presumption. And you must know that unless this is com-
plied with, and those transgressors sufficiently punished, I shall
dispose of my estate, and remove my family to some other part of
the world, where I may better support my dignity. And I do not

doubt but my example will be soon followed by many others, who have proper Spirit and Notions of their importance."

In 1766 these newspaper writers knew who each other was, and "R. R." was identified as one who earlier had posed as "A Man of Principle." In a Parthian shot, "Dikephilos" mockingly expressed his hope that when this agitated and unrealistic gentleman and his "other important great ones" remove from Virginia "to support your dignity elsewhere, you may, by retaining and cultivating your principles, make no trifling figure in the eloquence of Billingsgate."[40]

When the House of Burgesses convened at this time, it also heard petitions from William Stark, William Rind, Purdie & Dixon, and Robert Miller "to be publick Printer." In the balloting, Rind won with fifty-three votes and Purdie & Dixon came last with a mere ten. Rind got the appointment, but on November 27, Purdie & Dixon announced in their paper that the loss of the public business to their rival might deprive them of their principal support if not totally destroy their existence. "Now as we have reason to believe it the almost universal desire of the country that there should be two presses maintained," Purdie & Dixon are determined to continue their newspaper if friends will support it, and on their part guarantee to keep it *The Virginia Gazette*, "not only in name but in substance." These printers kept their word, and their newspaper or its successor lasted until 1781.[41]

Powerful support for a free press existed at this time. It came not alone from the burgesses of the upper counties, who had brought in William Rind and written generously for Purdie & Dixon, but from an increasing number of thoughtful Tidewater gentlemen. As early as August, "Philanthropos" expressed his satisfaction in the Williamsburg press as "one of the principal handmaids of liberty," though he deplored the pressure on both houses to print material traducing and threatening men of patriotic spirit. In September, "Dikephilos" conceded that the freedom of the press was often abused and begged all those who wrote for the public "to be *cool, deliberate, just* and *decent*. The consequences of men of ability writing with passion, and indecency, must I humbly apprehend, shock and give sincere concern to every well-wisher to society."[42]

The newspaper exchanges of this year undeniably contributed to the formation of a better-informed and more enlightened public in the Old Dominion. The attitude of the rank and file was quite properly best expressed in an unsigned letter "To Mr. Purdie" telling of "the real satisfaction I enjoy in the liberty of the press; which from its first erection here by Mr. Parks, in 1735, was in great measure shut up until you and Mr. Rind made it a free channel, whereby men may convey their sentiments for the amusement, instruction, or information of their fellow subjects." The recent publication of several fine instructive pieces hitherto held back by gentlemen, he observed, has been one of the good effects of this freedom, which he dated from March 7, 1766, and the coming of William Rind in May. For the first time in the colony's history, men in office have been freely censured for their acts and the Grand Jury has lately come out strongly for liberty. Therefore this citizen urged the Assembly to divide the public printing equally between the two houses in order to ensure the publication of two newspapers.[43]

The year came to a close with Rind's reprinting of David Hume's well-known essay "Of the Liberty of the Press," and on New Year's Day there appeared some animadversions by "Marcus Curtius" on Purdie & Dixon's *Virginia Gazette*, celebrating the superiority of its leading contributor over his detractors in recent issues of Rind's paper:

> Dikephilos, thy worth sublime
> (Too great for feeble prose or rhyme)
> Exceeds my best Endeavor.[44]

V

THE IMPORTANCE OF THE TRIVIAL

This recital has dealt with the seeming triviality of a colonial murder case, the sensationalism of which has had no more meaning to historians than the equally gory Lizzie Borden case. But we have also seen that when the Routledge-Chiswell affair is examined in connection with other contemporary events of historical importance in Virginia during 1765–66, Colonel John Chiswell's homicidal act assumes an unrecognized significance.

This unfortunate incident furnishes posterity with a richly detailed case history revealing just exactly how an aristocracy acts when it is in danger or merely threatened. Among themselves aristocrats will be democratic, as in a gentlemen's club; they will admit and even condemn ungentlemanly conduct by one of their own members, but they always present a solid, silent phalanx to the rest of society whether their cause be just or unjust, good or bad, right or wrong.

Colonel John Chiswell had murdered Robert Routledge[45] in "a strange fit of aristocratic insolence." He was in many respects both an able planter and a revolting cad. Willingly or unwillingly, he had cheated justice. Because of his high place among the Virginia gentry, his friends and intimate connections came at once to his defense. William Byrd III and Presley Thornton played several different roles in the scandal: as his business associates and creditors, as executors under Chiswell's will, as Council members, as judges of the General Court, and as the dispensers of apparently unwarranted special privilege to an accused murderer. A friendly King's Attorney conveniently left Cumberland County the morning the Examining Court convened; and in Williamsburg people doubted whether the Attorney General, a connection of Chiswell by marriage, would press the prosecution of the accused in the public interest. Nasty rumors of tampering with witnesses circulated and were never specifically denied in the press. Behind the scenes some other connections undertook to exercise political influence while still others wrote in the gentleman's cause after the newly freed press aired all of this, for what the Chiswell supporters had not foreseen was the opening of the columns of two rival newspapers at precisely this juncture.

What disturbed those of the gentry who were not involved in these proceedings or who were disgusted with the arrogance of the accused and what aroused the greatest public concern was the unabashed attempt of the Williamsburg clique to clear Chiswell at the very time of the uncovering of his son-in-law's largesse with the public currency. (Nobody knew until the 1950s that Byrd and the Lead Mine owed the most money to the Robinson estate!) Tensions over the Stamp Act and these happenings had divided the gentry into factions, as Governor Fauquier, John Camm, Councilor Carter, William Nelson, Robert Hartswell, Jonathan Boucher, and all the gazette writers frankly recognized.[46]

Chiswell's friends had counted upon being able to maintain the secrecy of action usual in the Old Dominion when such situations arose. We know how seriously—and fatally—they miscalculated. The planters of the "upper counties" insisted upon a properly conducted trial of the man who had run through one of their most popular associates, for they feared the opposite. They argued the point in the newspapers and they threatened trouble should justice falter. Thus the fears of the few that there would be no justice were communicated to all the inhabitants, and all Virginia worried about the outcome. John Chiswell's death ended further speculation on the subject.

The more that serious men read and thought about the deficiences in the legal system and the breaches of public trust in the House of Burgesses, the more some of them began to question the governing of Virginia in the interests of a class by a small faction of that class. These patriots understood very well the danger for the Old Dominion that lay in the perpetuation of a corrupt aristocracy like the one in England they censured. And had not the people of Virginia been told about public and private problems, the factions of a social and political caste generated by rumor, resentment, and recriminations, not to suggest further abuses of political trust, might have become permanent. Fortunately it fell out that by a species of unvoiced reform, the minority of bad gentlemen was brought into line by the majority of good gentlemen. The signal was given by the election of most of the supporters of Patrick Henry's Stamp Act resolutions to the new Assembly that met in November 1766, and the failure of a number of the older ruling clique to win seats. Until the Burgesses met, there had truly been a factional crisis in the gentry, but thereafter they coalesced as all gentlemen realized what a fatal impasse they had so narrowly avoided. Thus united, the Virginia aristocracy could face British authority with confidence in 1775 knowing full well that the "middling and inferior sorts" would loyally follow them.

IX

Baths and Watering Places
of Colonial America*

Aristocracy in America attained its apogee between the close of the French and Indian War and the meeting at Philadelphia of the Second Continental Congress. These were the happy years when, socially and spiritually, as well as in trade and politics, the gentlefolk of all the colonies came to know and understand each other better, to exchange amenities, to intermarry, and to sense their common bonds in the leadership of a truly American society.

Indeed, the gentry merged into the first class-conscious segment of the American people. The force most consistently urging this coalition was the universal human desire to exhibit one's wealth and exclusiveness to the best advantage for the admiration and envy of all. No contributing cause of local origin, however, was more potent than the marked advance in the means of intercolonial communication effected in these same years. In travel there was unity. More and better roads, stage lines and regular packet sailings combined with wealth, gentility, love of sociability, desire to imitate the mother country, curiosity, and the more imperative demands of good health to produce on this continent for the first time an organized attempt at conspicuous display apart from the homes of gentlemen and ladies. In the familiar guise of taking the

---

* This essay is reprinted from the *William and Mary Quarterly,* 3d Series, III (April 1946), 152–181, with the permission of the Editor.

waters at a summer resort, gentle valetudinarians sought that prized conjunction—restoration of health and relief from *ennui*.

Seventeenth-century colonists being English were nearly as prone as those who stayed at home to seek out and try the efficacy of mineral springs. In this they were aided by the Indians whose faith in the recuperative powers of mineral waters was of long standing. Although Bostonians patronized Lynn Red Spring as early as 1669, and William Penn, to his surprise, discovered that "there are mineral waters which operate like *Barnet* and *North Hall* that are not two miles from Philadelphia," the great colonial vogue for spas did not set in until the last years of the old French War.[1]

I

Conservative Bostonians continued to resort to Lynn Spring or to take an occasional plunge in the Cold Bath on Cambridge Street until 1766, when a new enthusiasm swept over from Connecticut with almost as exciting a result as the religious fervor of previous decades.[2] Perhaps some in town, like the post rider, knew of the hamlet of Stafford on the highway from Worcester to Hartford. Some may even have heard of the annual pilgrimage thither of Mohegans and Narragansetts to join with townsfolk in a restoring draught. But not until 1765 did Stafford Springs make the front page of the Boston papers.

For months in 1765 a Mr. Field of Windsor, Connecticut, had been afflicted with an obstinate cutaneous complaint, and that summer, in desperation, he sought relief in the waters of Stafford.[3] The treatment was brief, the cure complete. Gratefully Mr. Field made public his story, and overnight Stafford, because of its "reputation of curing the gout, sterility, pulmonary, hysterics, etc.," became the Mecca of New England's hypochondriacs. It was, asserted Samuel Peters, "the New England *Bath*, where the sick and rich resort to prolong life and acquire the polite accomplishments."[4]

Stafford's sudden rise to fame and fortune inevitably spawned rivals. In 1766 Ebenezer Guy of Windham, searching for springs "more handy to come at," discovered one at Mansfield, Connecticut. In "An Account of Experiments on the Waters of a Mineral Spring at Mansfield, and other Springs," which he succeeded in having printed in New London and New Haven newspapers, he declared that the Mansfield water possessed "the same nature and quality as that of Stafford," and was in his opinion actually superior.[5] June of the next year saw the completion of "accommodations" at a mineral spring in Newton, Massachusetts.[6] At Boston on August 6, Jackson's Mineral Well opened to dispense water to drink on the premises or "to carry away" in quart bottles. Ten days later Jackson advertised his Mineral Bath, announcing that it "will be filled every morning by 6 o'clock, and the Persons who bathe take their Turns as they come. All the time the Company are bathing, the Water will be convey'd into the Bottom by Pipes, and flow'd over the Top at the Rate of a Hogshead in four Minutes; and after every Day's Use, it will be drawn off, and well cleans'd with hot Water." Any time after nine in the morning "the Hot Bath" can be filled "in a few Minutes." Backed by the guarantees of two leading practitioners that his waters "may be serviceable in many disorders," Jackson doubtless turned over many an honest shilling.[7]

These challenges to the supremacy of the Stafford Springs were in vain. All over eastern Massachusetts, and especially at Boston, there arose in 1766 such an unquenchable thirst for Connecticut waters that good old New England rum and smuggled Madeira seem to have been temporarily forgotten. Finally, in early August, Dr. Joseph Warren followed his patients to Stafford "to examine the Waters that have been lately so much the Subject of conversation, in order to ascertain their Qualities and Uses." "Such examinations (for which the Dr. is well qualified)," the *Boston Gazette*'s inspired account stated, "by proper Experiments, will enable him to determine in what Distempers they are proper or improper—by which Means all Persons resorting to the Spring will have an opportunity to consult him respecting their Distempers and the Use of the Waters. If the Dr. meets with proper encouragement at Stafford, perhaps it may induce him to stay some Time to promote so laudable a Design. It is to be feared that these Waters may do Mischief, unless some able Physician is on the Spot to give proper advice."[8]

Scant success attended Dr. Warren's efforts, but in September the *Boston Post-Boy* reprinted from the *Connecticut Courant* a letter of Thomas Clap on Stafford Springs. This president of Yale had visited them, and though his experience was too brief to warrant judgment, yet from conversation with physicians there, he declared himself satisfied that the spring possessed "the same excellent Properties and Virtues which are found in some of the Medicinal Waters in Europe." So many great and surprising cures have been effected at the spring that it merits public attention and care; but improper and indiscreet use by injudicious persons for every kind of distemper, declared Mr. Clap with all the dignity and unction customarily employed by college presidents in delivering *obiter dicta,* may produce as many harmful as salutary effects, or "at least may be the Occasion of much Mispense of Time and Money." To his well-meant eight-point program to maintain a "skilled physician" in residence at the spring for advice to "unskilled Multitudes" who had "'no opportunity of taking any Direction, but of those who are as inexperienced as themselves," local practitioners turned a deaf ear.[9]

It took a calculating "Itinerant Medicaster" from old England to glimpse the financial possibilities of the New England spa. After advertising in Connecticut and Massachusetts papers, Dr. Tudor, "Lately from London," took up his residence on June 23, 1767, at Stafford, where he was available for "Advice to those who choose to consult him in drinking the Water."[10] Dr. Tudor's immediate harvest of fees aroused the professional jealousy and fanned the smoldering nativism of a true Yankee patriot who, as "Democritus," devoted a column and a half of eighteenth-century wit in the *Boston News-Letter* to "the present Flux of Medical Assistance from England." What New England needed was a medical school to foster native physicians in a profession that could rise to the dignity of the law and the church. At that time quacks abounded like the locusts in Egypt. "Are you for repairing to the Fountain of Health at Stafford" he sarcastically inquired. "Behold one of the Royal Race of *Tudor's* will deal out Salvation to you in a copious Deluge, and secure you from future Decay . . . ; 'tis apparent from the great Use made of these novel Glyster-mongers, we were prodigiously deficient before their happy Introduction among us."[11]

To New Englanders this display of bad taste among physicians was of no great interest. The newspapers more nearly took the public pulse when they declared that "as it is become a pretty general Practice to drink Mineral Waters, it is a Pitty so salutary a Medicine should be used with so little Discretion as is often the case. If some Gentlemen of Note of the Faculty would be so generous as to prescribe some general Rules for Persons who use the Waters, it might be of great Service."[12] Naturally, no tangible result came from this call not only for socialized medicine but for free medical advice.

With their wonted fine disregard for clerical and medical warnings, the multitudes of sick and rich continued to flock to Stafford to drink and be seen. In answer to the needs of "the middling Sort" who did not own coaches and chairs, enterprising John Wood at the Sign of the Lamb in Boston inaugurated a stagecoach and wagon service to Stafford on May 20, 1767—"The Price Five Dollars each Person, with an Allowance of 20 lb. weight." As often thereafter as "a full Company" offered, Wood made the seventy-mile trip over the road to Stafford.[13] The popular taste for mineral waters proved no short-lived fad. As many or more people went to Stafford in 1768 as in previous years.

One of them submitted anonymously an account of his experiences to the *Boston Chronicle,* "because it is a bound paper and will probably be better preserved." He wisely held that the great virtue of drinking the waters and taking cold baths regularly lies in prevention rather than cure. Both are a pleasure for most people, but to secure the best results patients must give over sitting with wine after meals. Nor are the benefits merely the product of the imagination as some cynics have charged. They are known to have reduced the "inflammatory state of the tubercles" of persons afflicted with consumption.

This precursor of Boston's famous L Street Brownies urged on his readers the cultivation of "the Therapeutic Art" of cold bathing. "It is to be regretted," quoth he, "that the force of popular prejudice operates so strongly, as to render it almost impossible to persuade the generality of people that it is either safe or beneficial; such are their panics, from the bare mention of it, that a traveller would be easily induced to believe that a pack of mad dogs had been let loose, and that hydrophobia or dread of water, consequent upon the bite of that animal, was a universal disease." So

healthful and curative did he find the practice that he strongly
urged as well bathing in common water and sometimes in the sea,
according to the following rules:

1. The best hours for bathing are in the morning, or any
   time before dinner, or towards evening.
2. Bathing immediately after "great eating or drinking" is
   deleterious.
3. One must never bathe after exercise or overheating.
4. Best results are had from bathing two or three times a
   week, each followed by a vigorous rub-down.
5. One must never bathe while experiencing a fever or a
   cold.[14]

How Stafford Springs appeared to a gentleman of New England
in 1771 is portrayed with his customary pungency in the *Diary* of
John Adams. Work and anxiety had overtaxed the constitution of
the thirty-five-year-old attorney whose friends advised him to try
the Stafford waters for his nerves. "May 30 (Thursday) Mounted
my horse for Connecticut," he noted.[15] As he rode lesiurely along
in a depressed state, Adams encountered all manner of gratuitous
advice from partisans and enemies of the springs. When he
reached the little village of 1,300 souls on June 3, Adams, like
many another weary traveler, "Lodged at Colburn's, the first
house in Stafford. There I found one David Orcutt, who came from
Bridgewater thirty years ago, a relation of the Orcutts in Weymouth. He, I find, is also a great advocate for the spring."[16]

Happy among friends and refreshed from a good night's sleep,
the young lawyer awoke the following morning and rode over to
the springs. "One Child had built a small house within a few
yards of the spring, and there some of the lame and infirm keep.
The spring issues at the foot of a steep high hill, between a cluster
of rocks, very near the side of a river. The water is very clear, lim-
pid, and transparent; the rock and stones and earth, at the bottom
are tinged with a reddish yellow color, and so is the little wooden
gutter, that is placed at the mouth of the spring to carry the water
off; indeed the water communicates that color, which resembles
that of the rust of iron, to whatever object it washes. Mrs. Child

furnished me with a glass mug broken to pieces and puttied together again; and with that I drank plentifully of the waters; it has a taste of fair water with an infusion of some preparation of steel in it, which I have taken heretofore [in Boston]. . . ." We fancy that fastidious gentleman made as much of a face at the plebeian mug as he did from the "noctious" taste of its chalybeate contents.[17]

"They have built a shed over a little reservoir, made of wood, about three feet deep, and into that have conveyed the water from the Spring; and there the people bathe, wash, and plunge, for which Child has eight pence a time. I plunged twice, but the second time was superflous, and did me more hurt than good; it is very cold indeed."[18]

From Mrs. Child, Adams learned that the best place to lodge was a half a mile away at Green's on the Common. He immediately moved his saddlebags there and discovered to his delight that the proprietor was an old friend from Worcester. From his chamber he had "command of a great view" and admired the vista down the "spacious road laid out very wide" with the meeting-house at the end. The next day, the novelty had worn off and he contented himself with a laconic entry in his diary: "Wednesday. Rode to the spring; drank and plunged; dipped but once; sky cloudy." Although he was cheered and impressed by the arrival of Dr. McKinistry with Colonel and Mrs. Bassell of Taunton at his lodgings, Adams, homesick for his wife and farm, left for Braintree. In after years he recalled that "this journey was of use to me, whether the waters were or not."[19] The same probably holds true for all the other visitors to Stafford, many of whom came all the way from the West Indies.[20]

II

The fashionable afflictions of gout, phthisis, and the vapors decreed for the rich seaboard gentry by a benign Providence were balanced in true eighteenth-century fashion when a perverse Fate visited the chills and ague upon all the inhabitants of the back country. Under either dispensation, the waters were thought to

bring relief, and they probably did, though in the West where
springs were free doubtless a greater proportion of the people
indulged.

Typical of frontier resorts was Warm Springs in Standing Stone
Valley, near the Juniata River in Pennsylvania, visited by the Reverend Philip Fithian in August 1775: "I rode to them through the
wet Bushes five miles, quite alone." One spring was used for
drinking; in the other, a kind of "Basin," six feet by four, had been
scooped out and concealed by pine boughs to provide privacy for
bathers. Fithian was pleasantly surprised to find the waters neither
cold nor bad tasting. They proved efficacious "chiefly in Rheumatic
Cases," violent body pains, weakness and debility, but some of the
more credulous folk swore they also offered "an Asylum for all im-
patient Women in Cases of Barrenness." In the company present
were twenty-four "professedly indisposed," including "seven un-
married Virgins" of various ages. "It looks indeed like an *Infirmary*
or *Hospital*, Many of them are by no means in Health. They must,
in strong Belief at least, be indisposed or they would not submit to
the Inconveniences for any Length of Time which the Situation of
the Place makes necessary. It is quite in the Woods." Between the
springs and Huntington there was not a single house or patch of
cleared land. "They must carry all their Provisions and Supply
themselves. They live in low Cabbins built with Staffs and Boughs,
and dress their Dinners all at one great common Fire. The Men,
for Exercise, play at Quoits, hunt Deer, Turkeys, Pheasants, etc.
With all these Hardships, however, they live in Friendship, and
are steadily cheerful, conquering by Society the *Uneasiness* both of
Infirmity and Labour, and making them almost constantly pleasant."²¹ Was this "L'âge d'or de Pensilvanie" of which French philosophers dreamed? Fithian did not dream, he merely carved his
initials on some flat stones beside those of "the great Numbers who
have been there before" and rode on his way to more fashionable
springs.

Most renowned of the back-country watering places were those
of Virginia.²² Before the arrival of white men these had been
frequented and valued by the Indians. Pre-eminent among them
were the "Fam'd Warm Springs" in Berkeley County, whose
waters George Washington first sampled in March 1748.²³ The following August, Joseph Spangenberg and Matthew Reutz, Mora-

vian missionaries recently arrived from a visit to the noted six springs at Conococheague in Pennsylvania, found those at Berkeley even better, and exclaimed at the sight of "a warm spring and a cold spring rising so close together, that, being in one, you can reach the other."[24] Two years later, to Dr. Thomas Walker, the waters of the warm spring seemed "clear and warmer than new milk." The six "invalids" he found there were in all probability from nearby settlements, for until the removal of the "Gallic menace," attendance by any but the most bold at Berkeley Springs would have been precarious in the extreme.[25]

Lord Fairfax gave the springs and a tract of land to Virginia in 1756, "that those healing waters might be forever free to the publick, for the welfare of suffering humanity." The fall of Quebec ushered in a new era. As early as August 1761 that indefatigable traveler, Colonel George Washington, sought a fortnight's relief from rheumatic fever at Berkeley Warm Springs. After a single day's treatment he wrote: "I think my fevers a good deal abated though my pains grow rather worse and my sleep equally disturbed." Sick as he was, Washington sent back to the Reverend Charles Green an account of the spring and the difficult trip out: "We found of both sexes about 200 people at this place, full of all manner of diseases and complaints; some of which are much benefited, while others find no relief from the waters." Food in abundance is available, "but lodgings can be had on no terms but building for them. . . . Had we not succeeded in getting a tent and marquee at Winchester, we should have been in a most miserable situation here."[26] By 1767 conditions had improved sufficiently to allow Mrs. Washington to accompany her husband and Colonel Fairfax on a seventeen-day trip to try the waters.

In a vain attempt to relieve the epilepsy of little Patsy Custis in 1769, the Washingtons drove out in their chariot to spend nearly two months at Berkeley Springs. Although, as in 1767 , he occupied Mr. Mercer's house, the Colonel had to repair it when he arrived and, in addition, erect an arbor for Patsy's comfort. Arranging for supplies, procuring the services of a baker and blacksmith, and finding a green pasture for his horses absorbed much of Washington's time and good humor. The latter was completely exhausted the day before he planned to start home when the chariot "broke."[27]

Berkeley Warm Springs had become by 1770 a place of great resort where the planting aristocracy, Jefferson's sturdy yeomanry, and coonskin democrats of the backwoods mingled at the baths much as did all classes of England in the pump room at Bath. Community of complaints and symptoms is a powerful democratizing agent. Down from Pennsylvania came rheumatic Indian traders like George Croghan to consort with Virginia Fairfaxes, Prentices, and Nicholases, as well as young bucks like John Hatley Norton from London.[28] The result was something less than Arcadian, but on the whole it was exceedingly human and very American.

Berkeley Springs early acquired most of the salient attributes of a summer resort—features we regard as modern. At least so it seemed to the fastidious, young clergyman, Philip Fithian, who arrived there after ten weeks of itinerant preaching in the backwoods. His diary reveals that he was properly shocked yet intensely intrigued and suffused withal by a very human envy of those not bound by Presbyterian cloth.

Life at the Virginia spa offered to the young minister round after round of excitement. On the very evening of his arrival occurred a "fray between Mr. Fleming and Mr. Hall concerning an Account. Mr. Hall wrung Mr. Fleming's Nose. I took Lodging at Mrs. Baker's. Mr. Miller, an Aged, Rheumatic Invalid taken ill in the Bath." The next morning Fithian "Drank early and freely of the Waters." Out of the four hundred persons then crowding the limited facilities of the village, he discreetly made "several Acquaintances" among the gentry. He missed meeting Parson Allen who had left for Frederick, Maryland: "It is said he has been mobbed by the Ladies." Tongues wagged about Parson Wilmer being "the veriest Buck in Town."[29] These gay Anglican vicars certainly put sober Presbyterians to shame.

That night at one end of the "little bush Village" was held "a splendid Ball. At some Distance, and within hearing, a Methodist Preacher was haranguing the People. . . . In our dining Room Companies at Cards—Five and Forty, Whist, Alfours, Callico Betty, etc." Eschewing hellfire and corruption alike, Fithian "walked out among the Bushes" only to find to his consternation "Amusements in all shapes, and in high Degrees, are constantly taking Place among so promiscuous [a] Company." A dignified chat with Mr. Edward Biddle of the Philadelphia Biddles, on leave from Congress because of rheumatism, served for a time to divert the young man's mind from these masques and revels—but only for a time.[30]

After the preaching, cards, and the ball ended, "from twelve to four this Morning, soft and continual Serenades" took place before the "different Houses where the Ladies lodge. Several of the Company, among them the Parson," Fithian notes almost enviously, "were hearty." One Miss, "said to be possess'd of an Estate in Maryland of ten thousand pounds, Is Accused by the Bloods as imperious and haughty." More serious is the accusation against a young blood "for breaking, and in the Warmth of his Heart . . . entering the Lodging Room of buxom Kate." Fithian, the man, felt for this "unfortunate Scot. He was led to this immediately stimulated by a plentiful Use of these Vigor-giving Waters. He came to recruit his exhausted System. He was urged, he was compell'd by the irresistable Call of renewed Nature. But Breaking Houses," moralized Fithian the preacher, "is breaking the Peace."[31]

Further up the Shenandoah Valley were other springs, which, by the eve of the Revolution, had begun to attract many Virginians. Thomas Jefferson thought the waters of Augusta Warm Springs were the most efficacious of all. Philip Fithian bore him out.[32] Located fifty miles from Staunton in mountainous country where on some nights "we hear Wolves howling in a dismal manner," were the springs whose contents the young clergyman averred "smells and tasts strangely like the Washings of a foul Gun." This after all was the proof of a mineral spring. "I drank almost a Pint of the Water; could not drink more; I washed my Face and Hands, this was pleasant." The warm bath, about thirty feet in diameter and from three to four feet deep, was enclosed by a pentagonal stone wall. During the summer months "Two or three hundred is the common Resort."[33]

Notwithstanding the very great distance of the Virginia waters from "the lower settlements," the unpredictable difficulties and inconveniences of travel, and save at Berkeley, the "total want of accommodation for the sick," the annual attendance of colonists at the several back-country springs during July and August must have approximated two thousand. Even the authorities in Williamsburg sought to make the springs more accessible to the planters: with £900 raised by lottery a "good Coach-Road" had been built over the mountains to the springs of Augusta County by 1772. On it, exulted Fithian, "you may gallop a Horse, and not hurt him or yourself." And in October 1776 an act was passed establishing the town of "Bath" at Warm Springs in Berkeley County. Fifty acres were set aside to be divided into lots and streets on which "to build convenient houses for accommodating numbers of infirm persons, who frequent these springs yearly." One large spring was designated as "suitable for a bath." Management of the bath and buildings was placed under the superintendence of a board of trustees, among whom were Bryan Fairfax, Samuel and Warner Washington, and Thomas Hite. Annually came a host from Virginia, Maryland, and western Pennsylvania. There would have been a much greater concourse at the springs, one contemporary thought, but for "the little Leisure which the greatest Number of Americans yet have."[34]

III

In striking contrast with the bucolic simplicity of New England's spas at Stafford and Mansfield, and the almost "gothick" wildness of the surroundings of the frontier springs of Virginia, stood the more urbane watering places of the metropolis of the colonies and its environs. As the resort of opulent and cultivated gentry, the spas of the Philadelphia area took on a sophistication that provided the nearest American approach to Bath and Tunbridge Wells. What began in 1700 as a search for medicinal waters had by 1744 become a fashionable indulgence. This craze amused John Bartram who reported to Dr. John Fothergill of London: "We have several springs in our province, on which many people have bestowed a large income."[35]

Nor did this spending cease! City Friends were soon impressed with tales told by rural Quakers of the marvelous curative properties and the charming prospects of the Yellow Springs. Situated among the rolling hills of Chester County about thirty miles west of Philadelphia, these springs became such a popular resort of Quakers on the Sabbath that several meetings had to spread warnings against the practice upon their minutes.[36] Construction of a public road in 1750 made the waters easily accessible, and John

Bailey, former silversmith of the city, shrewdly catering to the well-known taste of Friends for comfort, erected a stone inn, which opened under the management of Robert Pritchard in August. More and more people now went to the springs, which were large and properly nauseous, being well impregnated with sulphur and iron. Before long their fame had spread throughout the Empire. Mine host in 1762 was James Martin who, the court record reveals, had kept the tavern for several years, amazingly enough, as much to "the Satisfaction and Ease of the neighbors in General" as to that of the "vast Concourses of People" who "dayly frequent the . . . Springs for their Health from Philadelphia and all Parts of the Country as well as from the West Indies and other foren parts."[37]

The Yellow Springs soon came within the realm of those who had no chaise or chair in which to make the trip, for in May 1763, a "neat Stage Waggon" driven by John Cobble left at six o'clock in the morning three days a week from in front of William Peter's house in Philadelphia to carry them to the resort. Even "the middling Sort" could afford a brief stay there when Jonathan Durell announced accommodations for twenty-two shillings a week. Thereafter the stage ran regularly during the summer seasons, with tri-weekly departures from the Indian Queen Tavern. Then as now people lost their valuables on these excursions, like the gentleman who advertised in 1767: "Dropt, a few days ago, between the Yellow Springs and this City, the outside Case of a plain gold Watch, lined with Pink coloured Sattin, and the Picture of Mr. Pitt, in the inside of the Case."[38]

Like many resorts of a later day, despite constant patronage by a large number of people, the Yellow Springs did not prove profitable. Management of the inn and baths changed hands four times between 1750 and 1753, and frequently thereafter. In 1770, and again the next year, John Bailey advertised his 150-acre tract, buildings and springs for sale at public auction, but since there were no takers, he had to operate the establishment himself. This was no simple task. At this time the buildings consisted of a two-story stone inn, with a piazza nine feet wide running the whole length, "so well accustomed as to receive 100–300 people daily in the summer season, besides the unhealthy and the infirm who come from all parts to take lodgings for weeks together for the benefits of the waters." There were, in addition, the three baths which could be filled or emptied in a short time, two of which were "enclosed by good houses" (thirty-five by eighteen feet), each equipped with a "drawing room," fireplace, sash windows and other "modern" embellishments. Outside were walks between rows of shade trees, and "romantick" paths in meadows or woods.[39]

Sometime in 1774 Dr. Samuel Kennedy took over the "noted INN at the Yellow Springs," which he operated as a sort of sanitarium for several years, advertising in a Philadelphia newspaper that "the Advantage of these Baths is well known to the public; an incontestable proof of which is the great concourse of people— from four to six hundred persons have convened there in one day in the summer season."[40]

After the Battle of Brandywine, September 11, 1777, the Continental wounded were moved to nearby Yellow Springs. Dr. Kennedy offered his facilities to the Congress for a permanent military hospital. For the accommodation of nearly 1,300 soldiers, an appropriation was made for the addition of a building ninety feet long to those already existing. Thus until the death of Dr. Kennedy in 1778, this pleasure and health resort played the same role in the Revolution, as have other spas in later wars, as a refuge and retreat for the sick and wounded.[41]

Twenty miles northeast of Philadelphia on the road to Trenton and New York lay the little hamlet of Bristol, consisting of about fifty houses. Half a mile away flowed a mineral spring that had attracted invalids as early as 1720, but the generally unhealthy climate along this part of the Delaware with its swampy shores had discouraged any large resort to the spring before 1760, when the first baths were built. E'er long, however, so many visitors assembled to drink the waters that the Borough Council in 1769 had to rule that those not there on legitimate business would be arrested unless they dispersed at the order of the constable or the bath-keeper.[42]

The rise of Bristol to supremacy among colonial spas may be ascribed in large measure to its genuine chalybeate springs, its nearness to Philadelphia, and its location in a settled community, but most of all to the promoting proclivities of a native son. Dr. John De Normandie was a well-qualified physician who, besides acquiring a lucrative practice at the spring, studied scientifically the effects of the waters in the treatment of various disorders. He embodied his conclusions in a letter to Dr. Thomas Bond, vice-president of the American Philosophical Society, which was read at its September meeting in 1768. Dr. De Normandie described his experiments to determine, by chemical analysis, if the waters of Bristol were impregnated with the same "properties" as those of European watering places. They proved to have "principles similar to those of the much-celebrated waters of the German Spa, with which they likewise agree in the effects which follow immediately upon drinking them; such as quickening the pulse, exciting an agreeable warmth in the stomach, promoting the appetite, and *occasioning a flow of spirits and a greater degree of chearfulness.*" [43]

Clinical observation, asserted Dr. De Normandie, is far more reliable than mere chemical analysis in the determination of the medicinal value of mineral springs, and in a second letter to Dr. Bond, which the Society listened to in October 1769, he discussed not only nine new chemical experiments but presented careful detailed case histories of ten patients treated at Bristol. [44] These studies convinced him that the waters many be "safely and successfully drank in many cases," being particularly efficacious for such common eighteenth-century complaints as ulcers, scrofula, and jaundice; and of measurable benefit in relieving rheumatism. They are indeed a specific for many ailments, "but in no cases have their good effects been more evident or remarkable, than in a depraved and debilitated state of the organs of digestion, arising from inactivity and a sedentary life, . . . or from excessive and free living." Impressed by these letters, the Philosophical Society had them published immediately in the *Pennsylvania Journal* and voted their inclusion with the "Medical Papers" in the first volume of their *Transactions*, which was published in 1771. The powerful backing of the medical profession, the leading body of savants, and the press were vouchsafed to no other American watering place. Such publicity bore fruit by attracting to Bristol many persons from the South and the West Indies suffering from "a relaxed state of the solid parts of the human body, brought on by living in warm climates." [45]

Dr. De Normandie devised other means than those medicinal to improve Bristol as a summer resort. Through his efforts, miasmic swamps near the town were drained and townsfolk were urged to make their homes neat, clean, and "suitable" for the accommodation of gentle guests. Visiting Quakers might attend nearby meetings or cross the river to Burlington. But since many southern visitors were Anglicans, this "Gentleman of Fortune, character and great public usefulness," persuaded his fellow vestrymen to enlarge the Bristol Episcopal Church in 1772. The Reverend Richard Peters, rector of Christ Church and St. Peter's in Philadelphia, informed the Bishop of London that "we have this Summer alternatively supplied for them, as there is in the Summer season a great resort of strangers of Distinction to that Town on account of a mineral Spring and Bath there." In October, Dr. De Normandie sailed for England with another physician to tour the watering places, where, doubtless, they hoped to pick up a few new ideas.[46]

Bristol was better equipped to care for summer visitors than any other colonial watering place. Financed by Dr. John De Normandie's ample fortune, the proprietors, John Priestly and Charles Bassonet, advertised in 1769 that the spring had been cleaned out, a convenient new bath and house had been erected over it, and that they were prepared to "provide proper lodgings" in Bristol for visitors at a reasonable rate. In addition a resident physician will keep "an exact register" of the treatment of each patient who drinks the waters and will offer his advice to the poor gratis. By the next year these facilities became inadequate and a construction program was undertaken, which, by 1772, resulted in an elaborate plant.[47] Widely traveled Dr. Benjamin Rush of the Medical School in Philadelphia was so impressed with Bristol Spring that he described it in detail in his popular therapeutic pamphlet. But let him tell us about it with his usual vividness.

The buildings of "Bristol Baths and Chalybeate Wells" consist of a room forty-four feet long, "compleatly furnished between the pump-room and room for bathing: here whilst the baths are preparing, and during the time of drinking the waters every hour, the company amuse themselves in a manner the most agreeable, and have an opportunity of walking (an exercise, during that time, essentially necessary) free from every inconvenience, arising from an exposure to the open air.

"To the northward, and adjoining this long room, is the pump, inclosed in lattice work, and so situated, that the drinkers are immediately supplied with the water through a small door made in the partition, between the long room and the pump, by a person attending for that purpose.

"The Baths are in separate rooms, at the south end of the long room, one for ladies and another for gentlemen, with dressing apartments to each of them. They contain four hundred and fifty gallons each, and are filled in five minutes, by pumps immediately over the principal spring, which, by the most exact observation, are found to yield one hogshead of water in five minutes. After bathing, the water is let off, and conveyed away by pipes under ground, so that every bather may be supplied with fresh water, without being delayed but a few minutes."[48]

Everything joined, opined the physician, to make Bristol an outstanding spa. Since the Philadelphia-New York stage passed through town daily except Sunday and the post riders brought the mail three times a week, sick persons have "an opportunity of corresponding with their friends throughout the continent." The highways in the neighborhood are among the best in America and an especially fine road for pleasure parties in chaises or on horseback winds along the banks of the Neshaminy Creek for ten miles. And not the least among many fine features of the resort are good meats, vegetables, poultry, and reasonably priced lodgings. If all this be true, one suspects even that querulous hypochondriac Mr. Matthew Bramble would have enjoyed life at Bristol.[49]

Spurred by this come-to-Pennsylvania "literature," gentlemen and ladies from distant colonies sought in ever increasing numbers to recover their health and enjoy agreeable society at Bristol Spring. Not all were so fortunate as the New York Verplancks: in 1774 Mr. Joseph Galloway, who dwelt on his beautiful estate at Trevose only a few miles from Bristol, wrote to Samuel Verplanck expressing the hope that he would "make Trevose the Place of your residence during your stay, and will not think of taking Lodgings at Bristol. You may have the benefit of the waters without injury derived from the heat or air of that place." The Verplancks, it seems, were among the first of the New York families who perennially discover the charms of Bucks County.[50]

The popular craze for taking mineral waters that infected the colonies after 1760 was nowhere so widespread as at Philadelphia. Within the city and its suburbs, old springs became popular again and new ones were discovered. From the time of William Penn, citizens were accustomed to having physicians prescribe a course of waters at a chalybeate spring in the Northern Liberties, but it suddenly achieved notoriety in 1761 when the Protestant ministers petitioned Governor James Hamilton to prohibit a lottery "for erecting public Gardens with Baths and Bagnios among us." "Were a hot and Cold Bath necessary for the Health of the Inhabitants of the City, they might at a small Expence be added to the Hospital," said the clergy, but a stop must be put to the people's "immoderate and growing Fondness for Pleasure, Luxury, Gaming, Dissipation, and their concomitant Vices." This definitely ended the first attempt to give Philadelphia a combined Spa and Vauxhall.[51]

Conditions had altered somewhat four years later, for Dr. John Kearsley, architect and vestryman of Christ Church, was not opposed when he erected a bath house over this spring, and those townsfolk who had neither the time nor money to frequent Bristol or Yellow Springs flocked to it. The neighborhood came appropriately enough to be known as Bath-Town.[52] John White and his wife, "living at the New Bath," advertised that they could accommodate ladies and gentlemen "with Breakfasting, or the best of Tea, Coffee and Chocolate, with plenty of *Good Cream* etc. which Articles may be had also in the Afternoon." Almost as an afterthought, White added that he would attend those who desired to go into the bath "by furnishing them with Brushes and proper Towels."[53] Somewhat grandiloquently known as "The Rose of Bath," this establishment was faced with competition the next year from William Johnson's tearoom that opened across the street and became famous for its cheesecakes. White, not to be outdone, improved his bath by the installation of an "Engine" that would, he announced, fend off former complaints that the bath water was not changed often enough. He also improved a "fountain" spring nearby for mineral drinking water. "Those who've seen the famous Baths in England," he disingenuously claimed, "say these are quite equal to them." Season tickets at one pistole each might be obtained from several physicians who recommended the waters to patients, or at White's. Bath-Town's waters appear to have attracted mostly the "middling" and "inferior sorts" among Philadelphians. Perhaps because they did not possess the virtues of other springs, Dr. Rush omitted them from his pamphlet. White apparently failed about 1770, but "The Cold Baths in Bath Town" reopened in 1775 under the management of William Drewett Smith at whose "Medicinal Store" in town one might purchase season tickets.[54]

All Philadelphia was agog in the spring of 1773 with the news of the discovery of a mineral well on a vacant lot owned by John Lawrence, Esq., just across from the State House at Sixth and Chestnut streets. Twenty-one experiments performed by James Hutchinson, apothecary of the Hospital and Medical School, under the direction of gentlemen of the faculty headed by Dr. Benjamin Rush, yielded the conclusion that the waters, with their "slight faetid smell," exceeded any spring in the province in strength of chalybeate properties. Ex-Mayor Lawrence, to render the discovery more convenient and generally useful, arranged with "an indigent person" to attend at the well, deliver water "to the poor gratis," and others for a trifling consideration. Little did the pauper dream what a lucrative concession he had acquired. The citizenry crowded to the well as the word went around of "its usefulness to several afflicted persons." Pretty Sarah Eve, fiancée of Dr. Rush, was a consumptive, and one day in May she noted, "after tea, B. Rush, Miss Harper and I took a walk: Curiosity led us to the Mineral Point, and persuaded us to drink, or rather *taste* the water, which is excessively disagreeable, but at present is looked upon as an universal nostrum."[55]

Apparently the thronging of all and sundry to Lawrence's pump drew the attention of Dr. Benjamin Rush to the indiscriminate drinking of mineral waters by invalids for any kind of disorder and by healthy folk who had no need for purgatives, emetics, or restoratives. Before the American Philosophical Society in June 1773 he read a paper, published by James Humphrey the next month as *Experiments and Observations on the Mineral Waters of Philadelphia, Abington and Bristol,* which described the particular maladies these waters might relieve or cure with directions for their use. His concluding admonition was eminently sound: "Mineral waters, like most of our medicines, are only substitutes for temperance and exercise in chronic diseases. An angel must descend from Heaven, and trouble these chalybeate pools, before we can expect any extraordinary effect from their use alone."[56]

On a manuscript copy of Dr. Rush's treatise, Vice Provost Alison of the College wrote the following illuminating comments about John Lawrence's pump: "The water lost its virtue within a few months after investigation owing to the contents of a neighboring necessary. The well being exhausted on account of the quantity of it drunk, it was found the well communicated with the necessary which gave the smell and sediment."[57]

The popularity of Yellow and Bristol springs inspired an avid search for mineral waters in the countryside around Philadelphia. At Gloucester, New Jersey, in 1767, Hugh Jones opened a lodging house near the Philadelphia ferry because the previous season "many persons were prevented from going regularly to the springs, night and morning as prescribed," by the absence of adequate accommodations. Water from this spring was also bottled for sale in Philadelphia. Not far from Mount Holly, on the Rancocas Creek was a "Plantation well known by the name of Spaw" with an "extraordinary" spring, "none to exceed it perhaps in America," which came up for sale in 1771.[58]

In Bucks County another spa was the much frequented "Abington Mineral Spring" situated at Moreland near the junction of the York and Horsham roads, fourteen miles from Philadelphia. First owned by Thomas Hallowell in 1768, it developed into a popular resort under the direction of enterprising Thomas French who erected a bath in 1772 and advertised in the Philadelphia papers, with some show of reason, that his location was one of the healthiest in the province. "At a Moderate Expense," he offered comfortable lodgings to those who came to drink his waters, which were "recommended by some of our best physicians" as superior to any within thirty miles of the city. Abington Spring received a puff in Dr. Rush's pamphlet, and its waters enjoyed a local reputation down to the present century.[59]

IV

The large part played by spas and baths in the lives of the peripatetic colonial aristocrats is picturesquely set forth in the diary of Elizabeth Drinker. She and her husband Henry, who came of the city's oldest family, stood at the top of Philadelphia's social ladder and enjoyed a very wide acquaintance among the leading families of the middle colonies. During a visit to New York in September 1769, they rose early, crossed over to Brooklyn on the ferry, and breakfasted with Henry and Hannah Haydock. Next the Drinkers proceeded "to Graves end and down to the Beach or sea Shore. H. D. went into the Surf, and then [we] Return'd and Dined at Garrett Williamson's. From there Rid to New Utrecht, walk'd down to the Sea Shore, and return'd thro' Flat Bush to New York, crossing the East River by Moonlight." Three days later they went with a large party to the beach, "where we all opposed H. D.'s going into the Surf, it being very high," fearing "the undersucking of the Waves."[60] At this time New Yorkers, like John Taylor, having no mineral springs to exploit, sought the custom of gentle families "who may come either from the West Indies, Carolina, Virginia, or elsewhere, for their health or amusements," by tame salt water bathing in an enclosure on the banks of the Hudson.[61] How this sporting Quaker couple from Philadelphia must have dazzled them.

The diary is silent concerning the six weeks the family spent at the Widow Merriott's in Bristol during the summer of 1770, but offers an intimate account of the following season at the same watering place. Upon arriving on June 22, Elizabeth Drinker and her son dined at John Kidd's, and went to the bath in the evening. Among the boarders then staying at Widow Merriott's were Parson Carter and his wife, the Reverend Richard Peters, Mrs. David Hall (whose husband published the *Pennsylvania Gazette*), David Hall, Jr., Mrs. Thomas Cash, and all their servants. Each day Elizabeth Drinker went to the bath, solely to drink, and in the early evening took a chaise ride with Molly Hall, or went for a horseback ride accompanied by Dr. De Normandie's lad, Amos, up the scenic Neshaminy Drive. Society at Bristol was lively and agreeable in the extreme, and what with the coming and going of such personages as George Morgan, Collinson Read, the Walns and Mifflins of Philadelphia, the Boudinots, Coxes, Hoffmans, Serles, and Tilghmans from the neighboring colonies of New York, New Jersey, and Maryland, one could easily postpone going into the cold bath.[62]

The arrival of the family physician, Dr. John Redman, on June 28, and of Henry Drinker the next day, probably goaded Elizabeth

into taking the plunge: "June 30, . . . S. Merriott, Dr., Molly
Hall, Anna Humber and Self, went this afternoon into the Bath. I
found the shock much greater than I expected!" The next day Elizabeth, full of good resolutions, drove over to the bath, but "Had
not the courage to go in." Following a pleasant horseback ride on
July 8, she determined to try again: "about noon I went into the
Bath, felt cleaver after it." So cold was the water that often she
"found a difficulty of Breathing for many hours after it." In view of
this, consider the shock to such visitors as Mr. McCollegh of Jamaica and Madam Fanny Osborn, her daughter, and maid who
came all the way from Santa Cruz.[63]

July seems to have been the busy month at Bristol. On the
night of the fourteenth, Widow Merriott lodged twenty-three people. One would like to know if her accommodations were considered "moderate"; during a six weeks' stay the expenses of Mrs.
Drinker and her son, with occasional weekend items for Henry, in
good Pennsylvania currency were:

To Washing	£20.16. 6
To Widow Merriott	16. 6. 0
Bathing (H. D.)	4. 0. 0
Bath Tickets	1. 0. 0
3 Bath Tickets (season)	9. 0. 0
"pd Peggy at Bath"	15. 0.11
Polly Campbell, maid	17. 0. 3
Polly's wagon fare	5. 0. 0
	£88. 3. 8

No wonder the number of people who came out to Bristol for the
day far exceeded the resident patrons at the bath.[64]

Henry and Elizabeth Drinker did not restrict themselves to the
waters of Bristol that summer. Arriving home on August 12 they
tarried only eleven days before setting out in their chaise on a
pleasure drive to Lancaster. They "foarded Schuykill, then went to
the Yellow Springs, din'd there, took a walk in the Meadow," and
then—"took a duck in the Bath." The modern reader of Elizabeth's
diary experiences almost as much of a shock as she did with this
"duck" when he discovers that it would be twenty-eight years

before this gentle Quakeress would again get "wet all over at once."[65]

In due time the other mineral springs received the accolade from these inveterate Drinkers. In 1772, though they knew "Numbers of People resort there," Elizabeth and Henry contented themselves by purchasing Gloucester Spring water in town at four pence a bottle. But on June 27 they drove out the old York Road to visit Moses Shepherd who lived opposite French's Abington Spring and Bath. "The Water tastes pretty strong," observed Elizabeth. And of course she joined the host of Front Street gentry who consorted with members of the Assembly at John Lawrence's down-town well. On at least four different occasions during its brief ascendancy in 1773, Elizabeth, accompanied by Molly Foulk, Betty Jarvis, or other neighbors, arose and drank at the "pump" between six and seven o'clock in the morning and thought she felt better for it. This summer the Drinkers made only a three-day excursion to Bristol Spring. They called on Dr. Cadwalader Evans who was dying, and there encountered their friend Charles Thomson, "the Sam Adams of Philadelphia."[66]

The last jaunt this Quaker couple indulged in before hostilities ended civilian travel was to Black Point on the Shrewsbury River in East New Jersey, four months after the Declaration of Independence. They and some other Quakers, including the Rhoadses and Wisters, lodged at John Hartshorn's and passed an enjoyable week at the seashore. Henry Drinker bathed in the salt water nearly every day from the Wisters' bath house. Salt water to Elizabeth seemed fit only for drinking: "I drank nearly a pint of it, which operated largely and speedily." There was in the air and surroundings of Black Point a subtle something that impelled visitors to express their feelings in verse. When Governor John Penn and a party were there in 1775, Joseph Stansbury, soon to be the Bard of Loyalism, sententiously sang the praises of "Shrewsbury Vale," notwithstanding the fact that there "Musquetoes and Sand-Flies abound." But it remained for Elizabeth Drinker to pen in playful punning rhyme the valedictory of the baths and spas of colonial America:

"Lines verging somewhat on the Bath(os) but intended as a tribute of gratitude to our Landlord B. Wister, for his kindness in building us a Bath-house."

Hail! thou noblest of Landlords who'rt worthy to
      stand,
On a par, any *day*, with the *Knights* of the Land!
'Mongst the minions of monarchs, no man, surely
      hath
Half the claim to the title of *Knight of the Bath!*

Thee I hereby do dub, who to tub us hast
      deigned,
And cry hail to the man who his favors has
      rained,
On a house that had else been a great deal too
      dry,
Though containing of *Drinkers* a dozen or nigh.

Not a step shall we stir, not a ride shall we take,
But a feeling of thanks in our hearts shall awake,
For thou'st come like the spring, sung by Poets in
      Odes,
and thy showers refreshing hast shed on the
      Rhoads!

And each sultry day when emerged from the tub,
I sit down with friends to partake of a rub(ber);
My skin shall be cool which the heat else would
      blister,
And the pleasures of *whist* be made greater by
      Wister. [67]

V

What, then, may be said of the contribution of the watering places
to colonial life? [68]
   The ending of the American phase of the Seven Years' War released a series of forces that combined to bring the springs into
fashion. Though there had always been more intercolonial travel
than is usually recognized, 1760 witnessed the beginnings of the
annual seasonal migrations of individuals and families, which made

the aristocracy more urbane and more cosmopolitan than many of their English contemporaries. The colonial gentry were on the move.

For one hundred and fifty years Americans had worked hard; now it was time for those who could to play. For the first time sufficient wealth existed to provide extended leisure for the upper class; for the first time, too, tradesmen possessed funds ample enough for a brief trip to the springs. Life had become sufficiently flexible to allow them free exercise of their ingrained gregariousness and they eagerly joined in this first large-scale American experiment in recreation.

There were, of course, the springs themselves; their popularity is patent. Thousands of our forefathers visited them in the decade before the Revolution. Fear of the fevers and the desire to escape the humid heat of the southern and insular colonies literally drove planters, officials, and merchants northward and westward, while at the same time the noise and summer discomforts of burgeoning urban centers of the New England and middle colonies induced an exodus to the more salubrious air of the country.

No one would question the fact that there was some medicinal value in a course of the waters or even more therapeutic benefit to be derived from the mineral baths. Prominent physicians such as Lloyd, Warren, Rush, and De Normandie frankly warned the public not to expect too much. It is certain, however, that drinking mineral waters could not do as much harm as the copious doses of Bateman's Pectoral Drops, Daffy's Elixir, or Venice Treacle daily purchased in staggering quantities by members of all classes from apothecary shops. American springs, especially those of Virginia, proved equal or superior to those of Europe.[69] It is also true that colonial spas were cleaner, their surroundings more airy and natural, and in consequence, their benefits more lasting.

A trip to the springs indubitably produced more mental than physical benefits. In a letter to his friend David Hall in 1771, the great London printer William Strahan cogently abumbrated the theory of vacations: "I am very glad Mrs. Hall begins to pick up again. Going to Bristol Waters must be of great Service to her. The Country Air, the Waters themselves, but above all, the *Changing the Scene* must, at this time, be peculiarly serviceable to her."[70]

In New England, Virginia, and especially in Pennsylvania, the

rage for taking the waters approached that of contemporary England. If it be charged that in pursuit of the waters and their ac-
companying diversions, the colonists merely aped the mode of the
mother country, the answer is that Englishmen likewise borrowed
the custom from the Continent. Furthermore, Americans could
claim that society in their watering places was less extravagant and
less formal than at Bath or Spa. Rather than being imitators of
their English cousins by attending the mineral springs, the colonists were simply living in the manner of the eighteenth century.

These resorts proved a potent factor in promoting colonial
union and in nourishing nascent Americanism. They were the most
significant intercolonial meeting places. At the spas, representatives of the aristocracy of each colony met in person. In one week
at Bristol, the Quaker Drinkers consorted with Anglican clergy-
men, Presbyterian laymen, and gentlefolk from North Carolina,
Jamaica, and Santa Cruz. Around forest springs, at sophisticated
pump rooms, and in lodging houses occurred a noteworth mingling
of gentry, their retainers, and occasionally members of the middling and inferior sorts. Some came back year after year; some
made life-long friends; some began protracted correspondence;
some learned of common American interests, which inspired a
sense of class solidarity. Thus it may be confidently asserted that
these watering places provided a powerful solvent of provincialism
at a time when it was most needed. This portended mightily for
the future.

Notes

I. OPECHANCANOUGH OF VIRGINIA

1. There remain only literary sources for the reconstruction of the story of Opechancanough; recent excellent archaeological and anthropological studies, unfortunately, afford very little assistance with the problem. An exception to this statement is a very fine essay by Nancy O. Lurie, "The Indian Cultural Adjustment to European Civilization," in *Seventeenth Century America*, ed. James M. Smith (Chapel Hill, 1959), especially pp. 49–52.
2. William Strachey, *Historie of Travell into Virginia Britania* [1612], ed. Louis B. Wright and Virginia Freund, Hakluyt Society, 2d ser., CIII (1953), xviii, xx, 104–5, 108; *The Jamestown Voyages under the First Charter, 1606–1609*, ed. Philip Barbour, Hakluyt Society, 2d ser., CXXVI (1969), I, 134, 153–54; *Travels and Works of Captain John Smith*, ed. Edward Arber and A. G. Bradley (Edinburgh, 1910), I, 134–35; *Records of the Virginia Company of London*, ed. Susan M. Kingsbury (Washington, 1906–35), III, 17; David Beers Quinn, *England and the Discovery of North America, 1481–1620* (New York, 1974), 454–55.
3. Strachey, *Historie*, 104–5, also 15, 34, 56, 58, 91, 150.
4. "Their great King Opechancanow (that bloody Monster upon 100 years old)," in "A Perfect Description of Virginia" [1649], *Tracts and Other Papers, Relating . . . to the Colonies in North America*, ed. Peter Force (Washington, 1837), II, No. VIII, p. 7; Barbour, *Jamestown Voyages*, I, 142.
5. *The Luna Papers, 1559–1561*, ed. Herbert I. Priestly (Deland, 1928), II, 139; Clifford M. Lewis and Albert J. Loomie, *The Spanish Jesuit*

Mission in Virginia, 1570–1572 (Chapel Hill, 1953), 15–18. (Cited hereafter as *SJM*); Ribadeneyra, "Life of Borgia," 145; Martinez, "Relation," 155–56 (my italics), Sacchini, "History," 221, all in *SJM*. Fathers Lewis and Loomie printed most of the pertinent original sources, together with translations and an account of the venture. See a critical review by C. J. Bishko in *The Virginia Magazine of History and Biography*, LXXIII (1954), 214–15. With D. B. Quinn, we place more trust in the narrative of Martinez than do Fathers Lewis and Loomie.

6. Some Jesuit writers have held that the Indian lived in one of their Society's houses for "six or seven years," during which he was under its instruction. We believe that on his *first visit* to Spain the American was tutored by the Dominicans and did not live with the Jesuits until his *second visit* in 1566.

7. Oré, "Relation," 180, *SJM*.

8. Here we follow Father Sacchini, who seems to say that Don Luis and the Viceroy met in Spain. Sacchini, "History," 221, *SJM*; but *cf.* Rogel, "Relation," 118, and Carrera, "Relation," 131, in *SJM*.

9. Sacchini, "History," 221, *SJM*; Michael Kenny, *The Romance of the Floridas* (Milwaukee, 1953), 148 (for the opposition of the archbishop).

10. "Letters from Pedro Menéndez de Avilés to the King of Spain," Massachusetts Historical Society, *Proceedings*, 2d ser., VIII (1894), 433, 466–67; Bartolomé Barrientos, *Pedro Menéndez de Avilés: Founder of Florida* [Ms, 1567], translated by Anthony Kerrigan (Gainesville, 1965), 26; L. A. Vigneras, "A Spanish Discovery of North Carolina in 1566," *North Carolina Historical Review*, XLVI (1969), 408. The policy of "congregating" in Mexico is described by J. H. Parry in *The Spanish Seaborne Empire* (London, 1966), 188–89.

11. Felix Zubillaga, *La Florida. La Missión jesuitica y la colonización Española* (Rome, 1941), 349, n.9, for the royal order as cited by Lewis and Loomie, *SJM*, 21; Vigneras, "A Spanish Discovery," 402–3.

12. Extract from the "Register" of Diego de Camargo, recorder of the expedition, in the Archivo Generale de Indias, published by Vigneras, "A Spanish Discovery," 408; Gonzalo de Solís de Méras, *Pedro Menéndez de Avilés* [1567], ed. Jeanette T. Connor (Deland, 1923), 208–9.

13. Vigneras, "A Spanish Discovery," 411–13; Solís de Méras, *Menéndez*, 209; Cumarga's Report of the Expedition, Archivo Generale de Indias, Patronato 257, No. 3–2–4, pp. 4–12 (translation kindly supplied by Father Loomie).

14. Lewis and Loomie, 24–25; Rogel, "Relation," 118, Carrera, "Relation," 131, Martinez, "Relation," 156, Oré, "Relation," 180 (my italics), in *SJM*; David B. Quinn, *North America from the Earliest Discovery to First Settlements*, New American Nation Series (New York, 1977), 279.

15. Martinez, "Relation," *SJM*, 156.

16. Rogel, "Relatio," 11; Carrera, "Relation," 131–32, Ribadeneyra, "Life of Borgia," 145, in *SJM*.
17. Quinn, *North America*, 279n., citing letter in Eugenio Ruidias y Caravia, *La. Florida* (Madrid, 1896), II, 299–300; Letter of Quirós and Segura, 89, Carrera, "Relation," 131–33, in *SJM*.
18. After first meeting Powhatan in January 1606, Captain John Smith was able to write: "Their chief ruler is called Powhatan, and taketh his name of the principall place of dwelling called Powhatan. . . . His kingdom descendeth not to his sonnes nor children, but first to his brethren, whereof he hath 3, namely, Opitchapan [Itopatin], Opechancanough, and Catataugh; and after their decease to his sisters . . . but never to the heires of the males." Smith, *Travels and Works*, I, 22, 79, 81; Letter of Quirós and Segura, 89, Oré, "Relation," 180, Sacchini, "History," 222 (my italics), 227, n.10, in *SJM*.
19. Letter of Quirós, 92, Martinez, "Relation," 158, Oré, "Relation," 180, in *SJM*; Strachey, *Histoire*, 61.
20. Oré, "Relation," *SJM*, 181.
21. Rogel, "Relation," 109, Carrera, "Relation," 158–60, Oré, "Relation," 181, in *SJM*.
22. William R. Gerard, "The Tapahanek Dialect of Virginia," *American Anthropologist*, n.s., VI (1904), 314, 314n.
23. According to Father Rogel, who accompanied the Menéndez expedition of 1572, "the order of the Governor was to take the uncle of Don Luis, a principal chief of that region [*de prender a un cacique principal de aguella ribera*], as well as some leading Indians," and to hold them until Alonso was safely delivered to Menéndez. The *Adelantado* wanted the uncle because the murder of the priests took place in his country and therefore he shared the responsibility with Don Luis, whom Menéndez also wanted captured. This uncle was "*a* chief," not "*the* chief" or Powhatan in whose safe custody the boy lived several days' journey away on the Pamunkey River. The latter was "*the* chief" or Powhatan who had offered to yield his authority when Don Luis came back to his tribes in 1570. Rogel, "Relation," 130, 135, 139, Martinez, "Relation," 159, 161–62; Oré, "History," 183–85, in *SJM*.
24. Frank G. Speck dates the beginnings of the Powhatan empire at *ca.* 1570, the same year as the return of Don Luis. It began with the six original tribes dwelling within a twenty-five-mile radius of present-day Richmond. The Pamunkey outnumbered the other five tribes taken together, and the "empire" was founded upon conquest and the despotic exercise of personal power by the Powhatan. "Chapters on the Ethnology of the Powhatan Tribes of Virginia," Heye Foundation, *Indian Notes and Memoranda*, I (1919), 292–93; Ralph Hamor, *A True Discourse of the Present State of Virginia* [1615], facsimile edition (Richmond, 1957), 13; Strachey, *Historie*, 91; *Records Va. Co.*, III, 566.
25. Quinn, *England and the Discovery of Virginia*, 452–56; Strachey, *Historie*, 34, 38, 58, 91, 150; Smith, *Travels and Works*, I, 5, 18, 20,

71–72; Samuel Purchas, *Hakluytus Posthumus, or Purchas His Pilgrimes* (London, 1625), IV, 1513.

26. Smith, *Travels and Works*, I, 1 (my italics), li–lii; Barbour, *Jamestown Voyages*, I, 134.

27. There seems to be little doubt that Smith had read in Thomas Harriot's *A Brief and True Relation* (1588) about the astonishment of the Indians at Roanoke upon first seeing a compass and that, recalling this, the nimble-witted captive made good use of his own compass in December 1607. *The Roanoke Voyages: 1584–1590*, ed. David Beers Quinn, Hakluyt Society, 2d ser., CIV (1955), I, 375–76; Smith, *Travels and Works*, I, 15–16; II, 396.

28. Smith, *Travels and Works*, I, cxvii n.; II, 396; Barbour, *Jamestown Voyages*, I, 181–82.

29. Smith, *Travels and Works*, I, lxxxvi, 18–20, 22.

30. Smith, *Travels and Works*, I, xxx, 18–20, 22, 141–42; II, 458–59; Purchas, *Pilgrimes*, IV, 1723.

31. Strachey, *Historie*, 64–65.

32. Hamor, *True Discourse*, 6–10; *Journals of the House of Burgesses of Virginia, 1619–1659*, ed. H. R. McIlwaine (Richmond, 1915), 33.

33. Hamer, *True Discourse*, 10, 53–54.

34. Hamor, *True Discourse*, 10–11, 53–54, 64–68; Smith, *Travels and Works*, II, 514.

35. Powhatan also said to Hamor of the peace: "So as if the English offer me injury, my country is large enough, I will remove my selfe farther from you." Encroachments upon Indian lands were not a grave issue at this time. Hamor, *True Discourse*, 37–43 (p. 42, my italics); Smith, *Travels and Works*, II, 519; *Journal of the House of Burgesses, 1619–1659*, p. 33.

36. Governor Dale was apparently completely taken in at Pamunkey. He wrote on June 18: "Opachankano desired I would call him friend, and that he might call me so, saying he was a great Captaine, and therefore he loved mee, and that my friends should be his friends. So the bargaine was made. . . ." Hamor, *True Discourse*, 54, also 10–11, 59, 61–69, index under "Opechankano"; *cf.* Smith, *Travels and Works*, II, 514.

37. Hamor, *True Discourse*, 27, 54–55, 59–60.

38. Hamor, *True Discourse*, 2, 16 (my italics).

39. Hamor, *True Discourse*, 11–14, 56–57; *Records Va. Co.*, IV, 117.

40. Smith, *Travels and Works*, II, 528.

41. William Stith, *The History of the First Discovery and Settlement of Virginia* [Williamsburg, 1747] (facsimile edition, Spartanburg, 1965), 138–39; Purchas, *Pilgrimes*, IV, 1774; Hamor, *True Discourse*, 2.

42. Smith, *Travels and Works*, II, 527–28; *Records Va. Co.*, IV, 118.

43. When Captain Samuel Argall succeeded George Yeardley as governor in May 1617, he was dismayed to find at Jamestown "the Salvages as frequent in their houses as themselves, whereby they were become expert in our armes, and had a great many in their custodie and pos-

session; the Colonie dispersed all about planting Tobacco." Smith, *Travels and Works*, II, 528–29.

44. *Records Va. Co.*, IV, 118; for Ozinies, see Smith's Map of Virginia of 1612, *Travels and Works*, facing 384, 528–29.

45. Opechancanough's gift of a large tract of land at Weyanoke to Yeardley in 1617 is truly puzzling. Whether it was made before or after May 15, when Argall replaced him as governor, is not known. The act may have been merely a part of the Indian's elaborate scheme of dissimulation to lull the English into a false security. The chief probably hoped to turn the white men's expansion up the James River Valley in order to relieve the pressure on the Chickahominy Valley, where the tobacco planters were beginning to intrude. Deed Book, General Court, No. 1, p. 82 (now destroyed) in Robinson Transcripts, cited by Philip Alexander Bruce, *The Economic History of Virginia in the Seventeenth Century* (New York, 1895), I, 490–91; *Records Va. Co.*, IV, 118.

46. *Records Va. Co.*, III, 92; Smith, *Travels and Works*, II, 540.

47. Purchas, *Pilgrimes*, IV, 1774; Smith, *Travels and Works*, II, 539; *Records Va. Co.*, III, 73–74; Robert Beverley, *The History and Present State of Virginia*, ed. Louis B. Wright (Chapel Hill, 1947), 45.

48. Beverley, *History of Virginia*, 45; *Records Va. Co.*, III, 92.

49. Nicholas Ferrar and others of the London Company wrote to Argall about this time (August 28): "Wee cannott imagine why you should give us warninge that Opechankano and the Natives have given their Country to Mr. Rolfes Child [Thomas] and that they will reserve it from all others till he comes of yeares." Nor can we imagine what was going on, unless it was another of the Powhatan's many subterfuges, such as the Weyanoke gift. *Records Va. Co.*, III, 52–53, 79, 92–93, 247; Smith, *Travels and Works*, II, 538–39; Alexander Brown, *The First Republic of the United States* (Boston and New York, 1895), 279, 281.

50. *Records Va. Co.*, III, 107.

51. *Records Va. Co.*, III, 147 (my italics).

52. *Records Va. Co.*, III, 14.

53. *Records Va. Co.*, III, 71, 128; I, 19, 319.

54. *Records Va. Co.*, III, 157–58.

55. *Records Va. Co.*, II, 404–5; III, 244–45.

56. *Records Va. Co.*, III, 244–45.

57. *Records Va. Co.*, I, 310; III, 174–75, 242, 244–45; *Narratives of Early Virginia, 1606–1625*, ed. Lyon G. Tyler, Original Narratives of Early American History (New York, 1907), 274–75.

58. John Rolfe wrote to Sir Edwin Sandys in January 1620 of Robert Poole: "that Sir George would make him sure [secure] for telling any more false tales to Opechancano, if once he gott him into his power; at Pooles coming lately home we were of quite another opinion. For bringing the King's picture [James I's?] as a messenger from Opochankanough, we counted him a publique, and as it were a neutral

person, and so for not discontenting Opochankanough, with whom we now stand in termes of reconciliation, we thought it no wayes convenient to call Poole to accounte." The Company in London referred to "the dishonor he did to the Kings Picture" at a meeting on October 23, 1622, without further detail, but they seem to have meant Opechancanough. Evidently the Powhatan was using the interpreter successfully as a pawn in his game with the English. *Records Va. Co.*, III, 253; II, 115.

59. *Records Va. Co.*, III, 228.
60. *Records Va. Co.*, III, 102.
61. *Select Charters and other Documents Illustrative of American History, 1607–1775*, ed. William Macdonald (New York, 1899), pp. 2–3; Smith, *Travels and Works*, I, xxxvi; Barbour, *Jamestown Voyages*, I, 25, 43, 51–52.
62. Alexander Whitaker, *Good Newes from Virginia* (London, 1613), 25, 26–27.
63. Smith, *Travels and Works*, II, 562.
64. *Records Va. Co.*, III, 446–47 (my italics).
65. Compare Smith's discourse with the chieftain. *Records Va. Co.*, III 446, 451, 583–84 (Opechancanough's astronomy).
66. *Records Va. Co.*, III, 552; Smith, *Travels and Works*, II, 574.
67. Smith, *Travels and Works*, II, 564; for the influence of the priests on decisions for war, see Strachey, *Historie*, 104.
68. *Records Va. Co.*, III, 467.
69. *Records Va. Co.*, III, 469 (my italics).
70. Beverley, *History of Virginia*, 226.
71. *Records Va. Co.*, III, 707.
72. Governor Wyatt discussed further the advantages of the league for "both Parts": "the Savages as the weaker, under which they were safely sheltered and defended; to us, as being the easiest way then thought to pursue and advance our projects of building, planting and effecting their conversion by peaceable and faire meanes." *Records Va. Co.*, III, 549–50; IV, 178; Brown, *First Republic*, 417.
73. *Records Va. Co.*, II, 94–95 (my italics); III, 309.
74. In the London Company's official version of the "Massacre," published in 1622, Edward Waterhouse wrote: "It is since discovered, that the last Summer [1621] Opachancano practised with a King of the Eastern Shore (*no well wisher of his*) to furnish him with a store of poison (naturally growing in his country) for our destruction, which he absolutely refuseth. . . ." The King and five or six of his great men "offered to be ready to justifie against him." The records contain no further reference to this, which suggests that the men of Jamestown did not credit a charge made after the slaughter by enemies of Opechancanough. *Records Va. Co.*, III, 438, 556 (my italics).
75. *Records Va. Co.*, IV, 10.
76. *Records Va. Co.*, III, 550; Smith, *Travels and Works*, II, 561, 565, 575.

77. In Governor Yeardley's time, Opechancanough had sent word by his interpreter concerning Nemattanow, that "for his parte he could be contented his throte were Cutt." Smith, *Travels and Works*, II, 572; *Records Va. Co.*, IV, 11.
78. *Records Va. Co.*, IV, 11.
79. *Records Va. Co.*, III, 550 (my italics).
80. Following "Master Wimp," Smith stated in 1624: "The Prologue to this tragedy was occasioned by Nemattanow." To interpret *prologue* to mean *cause*, as some writers have done in this instance, is to ignore the course of events from 1613 to 1622. Master Wimp must have meant *pretext*—immediate cause—and the Council in Virginia so understood it in January 1623. Smith, *Travels and Works*, II, 572; *Records Va. Co.*, IV, 11.
81. *Records Va. Co.*, III, 549, 550, 554.
82. *Records Va. Co.*, III, 555, 612.
83. George Wyatt to his son, Sir Francis, undated, 1624, *William and Mary Quarterly*, 3d ser., XXXIV (1977), 116–17.
84. A few days after the "Massacre," a "savage" stole aboard the *Elizabeth*, which was lying at anchor up the Potomac, gave Captain Henry Spelman the news, and related further "that Opechancanough had plotted with his King and Countrey to betray them also, which they refused; but them of Wighcomoco at the mouth of the river had undertaken it." Smith, *Travels and Works*, II, 586; *Records Va. Co.*, III, 542, 551, 552–53.
85. Beverley, *History of Virginia*, 50; *Records Va. Co.*, III, 542.
86. One Virginian wrote to London: "Wee freely confesse, that the Countrey is not so good as the *Natives* are bad, whose barbarous Savagenesse needs more cultivation than the grounde it selfe, being more overspread with incivilitie and treachery, then with Bryers." *Records Va. Co.*, III, 549; II, 115; Smith, *Travels and Works*, II, 586–87, 591.
87. *Records Va. Co.*, III, 556–57.
88. Purchas, *Pilgrimes*, IV, 1810, 1813; *Records Va. Co.*, III, 666, 671–72, 673, 683, 704–7, 708; IV, 9, 10, 12; Brown, *First Republic*, 473; Beverley, *History of Virginia*, 54.
89. John Aylmer, *An Harborowe for Faithfull and True Subjects*, quoted in Carl Bridenbaugh, *Vexed and Troubled Englishmen, 1590–1642* (New York, 1968), 13; "Good News from Virginia" [1623], *Photostat Americana* (Massachusetts Historical Society), 2d ser., No. 105 (1936), or in *William and Mary Quarterly*, 3d ser., V (1948), 355, 356.
90. *Records Va. Co.*, III, 673.
91. Massachusetts Historical Society, *Collections*, 4th ser., IX (1871), 69–71; *Records Va. Co.*, IV, 6–9, 71; III, 672, 683.
93. *Records Va. Co.*, II, 483; III, 102; Brown, *First Republic*, 495, 514.
94. *Records Va. Co.*, IV., 105, 221–22.
95. The Virginia Company of London fully approved the local Council's resolution to use "all good courses" of policy "so farr as may stand with Justice and truth . . . yet are you to[o] worthie to use any false

dealing, and we desire that yor proceedinge may be so faire as may not only be free from the fault, but any just ground of suspicion." The report of June 11 from Jamestown had evidently reached London by August 6, the date of this letter. Opechancanough still held 19 prisoners on June 19. *Records Va. Co.*, IV, 89, 221–22, 232, 260–70; II, 486; Robert Rich, Earl of Warwick, to the Earl of Conway, as cited by Powell in *Virginia Magazine*, LXVI (1958), 62.

96. *Records Va. Co.*, II, 478; IV, 450–51; Brown, *First Republic*, 576–77.

97. Harvey's Declaration, *ca.* February 1624, Massachusetts Historical Society, *Collections*, 4th ser., IX (1871), 69–70, 71.

98. *Records Va. Co.*, IV, 507–8; Frederick Merk, *History of the Westward Movement* (New York, 1978), 89.

99. On August 12, a proclamation was issued to make known that "the Indians are not to be taken as friends but can be taken of them as if enemies in actual war." *Minutes of the Council and General Court of Virginia, 1622–1632, 1670–1676*, ed. H. R. McIlwaine (Richmond, 1924), 198, 484; Captain William Perse's Relation, August, 1629, Public Record Office (London), C. O. 1/5, pt. I, fol. 69, as printed in Edward D. Neill, *Virginia Carolorum* (Albany, 1866), 60–61.

100. *Minutes of the Council, 1622–1632*, pp. 480, 484; *Journals of the House of Burgesses*, 52, 53; William Waller Hening, *Statutes at Large of Virginia* (Philadelphia, 1823), I, 176, 177.

101. "A Perfect Description of Virginia" [1649], Force, *Tracts*, No. 8, p. 11; Beverley, *History of Virginia*, 60–61.

102. "He heard one Day a great Noise of the Treading of People about him; upon which he caused his Eye-lids to be lifted up; and finding that a Crowd of People were let in to see him, he called in high Indignation for the Governor; who being come, Opechancanough scornfully told him, That had it been his Fortune to take Sir William Berkeley Prisoner, he should not meanly have exposed him to a Show to the People." Beverley, *History of Virginia*, 60–62; *Winthrop's Journal*, ed. James K. Hosmer, Original Narratives of Early American History (New York, 1908), II, 167; *Mercurius Civicus*, May 22, 1645, reprinted in *Virginia Magazine*, LXV (1957), 85.

103. Beverley, *History of Virginia*, 60–62.

104. On August Herrman's map, *Virginia and Maryland* [ca. 1670] (London, 1673), between the James River above the falls and Appomattox River is marked:

> Here about Sir Will. Berkley,
> conquered and took Prisoner
> the great Indian Emperour,
> Abatschakia, after the
> Massacre _____ in Virginia,
> Ano. [no date]

Beverley, *History of Virginia*, 62; Hening, *Statutes at Large*, I,

323–26. See in general the reflections about Opechancanough by Mrs. Lurie, in *Seventeenth-Century America*, 50–52.
105. Joseph Alsop, *The Silent Earth* (London, 1965), 236.

II. THE OLD AND NEW SOCIETIES
OF THE DELAWARE VALLEY IN THE SEVENTEENTH CENTURY

1. *Publications of the Colonial Society of Massachusetts, Transactions 1947–1951*, XXXVIII (Boston, 1959), 50.
2. The *Oxford English Dictionary* defines *society* as "the living in association for the purposes of harmonious existence by a body of individuals in a more or less ordered community." *Society* is used in this essay in preference to *culture* which, as the social scientists employ it, represents a species of professional jargon that does violence to the original meaning of the word. Moreover, to the layman, *culture* obscures more than it clarifies.
3. Paul A. W. Wallace, *Indians in Pennsylvania* (Harrisburg, 1961, 16.
4. Peter M. Lindeström, *Geographia Americanae, with an Account of the Delaware Indians . . . 1654–1656* (Philadelphia, 1925); Wallace, *Indians*, 4, 17–83; and in general C. A. Weslager, *The Delaware Indians: A History* (New Brunswick, N.J., 1972).
5. Albert Cook Myers, *Narratives of Early Pennsylvania, West New Jersey, and Delaware, 1630–1707* (New York, 1912, 1959), 18–24; Thomas Campanius Holm, "A Short Description of the Province of New Sweden," Historical Society of Pennsylvania, *Memoirs*, III (1884), 158; Francis Jennings, ". . . The Susquehannock Indians in the Seventeenth Century," American Philosophical Society, *Proceedings*, 102 (1968), 15–21.
6. Paul A. W. Wallace, "Historic Indian Paths of Pennsylvania," *The Pennsylvania Magazine of History and Biography* (hereafter cited as PMHB), LXXVI (1952), 411–39, and folding map facing p. 438; Robert Proud, *History of Pennsylvania in North America* (Philadelphia, 1797), II, 253–55, for creeks and portages; and for the Indian paths of New Jersey, see Wheaton J. Lane, *From Indian Trail to Iron Horse* (Princeton, 1939), 6, 17 (map of trails), 24.
7. Amandus Johnson, ed., *Instructions for Johan Prinz, Governor of New Sweden* (Port Washington, N.Y., 1969), 136, 161, 177, 187–88.
8. Myers, ed., *Narratives of Pennyslvania*, 237–38.
9. C. A. Weslager, "Log Structures in New Sweden," *Delaware History*, V (1952), 77–89; *Archives of Maryland* (Baltimore, 1884), II, 224–26; H. Clay Reed, and George J. Miller, eds., *The Burlington Court Book . . . 1680–1709* (Washington, 1944), 11–12; Bartlett B. James and J. Franklin Jameson, eds., *Journal of Jasper Danckaerts, 1679–1680* (New York, 1913, 1959), 96, 97, 98, 101, 148; Myers, ed., *Narratives*

of Pennsylvania, 250; and the standard work of Harold R. Shurtleff, *The Log-Cabin Myth* (Gloucester, Mass., 1967).

10. *Instructions for Prinz*, 177; Myers, ed., *Narratives of Pennsylvania*, 107, 125; Edward B. O'Callaghan, and Berthold Fernow, eds., *Documents Relative to the Colonial History of the State of New York* (Albany, 1856–87), II, 9, 10, 16, 76, 210, 211; Colonial Society of Pennsylvania, *Records of the Court of New Castle on Delaware, 1676–1681* (Philadelphia, 1904), I, 81, 94–95; Jameson, ed., *Journal of Jasper Danckaerts*, 105.

11. Julius F. Sachse, ed., *Letters Relating to the Settlement of Germantown* (Philadelphia, 1903), 13.

12. Myers, ed., *Narratives of Pennsylvania*, 96, 97, 145, 157; *Instructions for Prinz*, 190–91, 258; Edward B. O'Callaghan, ed., *Laws and Ordinances of New Netherland, 1638–1674* (Albany, 1888), I, 307–9; *Calvert Papers*, Maryland Historical Society, *Fund Publications* (Baltimore, 1888), I, No. 28, p. 76.

13. Amandus Johnson, *The Swedish Settlements on the Delaware . . . 1638–1664* (Philadelphia, 1911), I, 335, 337, 339, 341; *N.Y. Col. Docs.*, I, 558; Francis Jennings, "The Indian Trade of the Susquehanna Valley," American Philosophical Society, *Proceedings*, 110 (1966), 406–24.

14. Myers, ed., *Narratives of Pennsylvania*, 103–4; Lindeström, "Map of New Sweden," *Geographia Americanae*, facing pp. 152, 156.

15. Lane, *Indian Trail*, 17 (map); *Instructions for Prinz*, 180–81; Johnson, *Swedish Settlements*, I, 435–36.

16. Myers, ed., *Narrative of Pennyslvania*, 139–41; Jameson, ed., *Journal of Jasper Danckaerts*, 127–28.

17. Myers, ed., *Narratives of Pennsylvania*, 151, 157; Staughton George, *Charter to William Penn and Laws of Pennsylvania . . . Preceded by the Duke of York's Laws . . .* (Harrisburg, 1879), 72, 151; Jameson, ed., *Journal of Jasper Danckaerts*, 130.

18. George Johnston, *History of Cecil County, Maryland* (Elkton, Md., 1881), 20, 33, 58, 64.

19. Percy G. Skrivers, "Seven Pioneers of the Eastern Shore," *Maryland Historical Magazine*, XV (1920), 396–409; *Dictionary of American Biography* (New York, 1943), VIII, 592–93; *Archives of Maryland*, III (1), 398–99; II, 199, 205², 330–31; Johnston, *Cecil County*, 72; *N.Y. Col. Docs.*, XII, 481, 483; Victor H. Paltsits ed., *Minutes of the Executive Council of the Province of New York . . . 1668–1673* (Albany, 1910), II, 556; Jameson, ed., *Journal of Jasper Danckaerts*, 113.

20. Herrman's map was published in facsimile by the John Carter Brown Library (Providence, 1948); Johnston, Cecil County, 58.

21. John L. Nickalls, ed., *The Journal of George Fox* (Cambridge, Eng., 1952), 618–20, 630–34.

22. William Edmundson, *Journal* (London, 1715), 93–96. For the use of this Quaker overland route by proselyting Anglicans in 1702, see

George Keith, *A Journal of Travels from New Hampshire to Caratuck* [N.C.] . . . (London, 1706).

23. *N.Y. Col. Docs.*, XII, 493.
24. *Instructions for Prinz*, 80; Weslager, *Delaware Indians*, 128.
25. *An Abstract or Abbreviation of some Few of the Many . . . Testimonys from the Inhabitants of New Jersey* . . . (London, 1681), 20, 27. (Italics mine.)
26. Myers, ed., *Narratives of Pennyslvania*, 344; Weslager, *Delaware Indians*, 139–40, 150, 161–63, 165; *Pennsylvania Archives* (Philadelphia, 1852), First Series, I, 47, 62, 64, 117.
27. *N.Y. Col. Docs.*, III, 345; Sachse, ed., *Germantown Letters*, 32; Marion D. Learned, *Francis Daniel Pastorius* (Philadelphia, 1908), 260–63; Marion Balderston, ed., *James Claypoole's Letter Book: London and Philadelphia, 1681–1684* (San Marino, Calif., 1967), 223.
28. *PMHB*, IV (1880), 315; Proud, *History of Pennsylvania*, I, 209n.; *Claypoole's Letter Book*, 223–24; Myers, ed., *Narratives of Pennsylvania*, 260.
29. *New Castle Court Records*, I, 136; Jameson, ed., *Journal of Jasper Danckaerts*, 115, 134; Edward B. O'Callaghan, ed., *Documentary History of the State of New York* (Albany, 1849), I, 98.
30. Myers, ed., *Narratives of Pennsylvania*, 191–92.
31. Proud, *History of Pennsylvania*, I, 153; George Scot, *A Model of the Government of the Province of New Jersey* (Edinburgh, 1685), 70–71; Myers, ed., *Narratives of Pennsylvania*, 191.
32. Myers, ed., *Narratives of Pennsylvania*, 199–216, 257–58, 317, 325; Rufus Jones, *The Quakers in the American Colonies* (New York, 1966), 522; William Penn to Robert Harley, *ca.* 1701, Historical Manuscripts Commission, *Fifteenth Report*, Appendix, Part LV, "Duke of Portland MSS" (London, 1897), 32.
33. Myers, ed., *Narratives of Pennsylvania*, 191, 232.
34. Jones, *Quakers*, 522; Sachse, ed., *Germantown Letters*, 35.
35. *PMHB*, IV (1880), 196; Myers, ed., *Narratives of Pennsylvania*, 332; Wayne Andrews, ed., "The Travel Diary of Dr. Benjamin Bullivant," *The New-York Historical Society Quarterly*, XL (1956), 55–73.
36. Myers, ed., *Narratives of Pennsylvania*, 320–21, 332–33. There is a copy of John Worlidge's rare map in the Library of Congress.
37. Colonial Society of Pennsylvania, *Records of the Court of Chester County, Pennsylvania, 1681–1697* (Philadelphia, 1910), 78.
38. *Bucks County Court Records*, 88; *PMHB*, IV (1880), 20; XXIX (1905), 102, 103; Myers, ed., *Narratives of Pennsylvania*, 267, 271; Sachse, ed., *Germantown Letters*, 26, 32.
39. Myers, ed., *Narratives of Pennsylvania*, 269; *PMHB*, IV (1880), 194–95; Bullivant, "Travel Diary," 69–70.
40. Thomas Budd, *Good Order Established in Pennsylvania & New Jersey* . . . (Philadelphia, 1685), 8; Sachse, ed., *Germantown Letters*, 32.
41. *PMHB*, IV (1880), 194, 197–98, 200–201; Bullivant, "Travel Diary,"

67–70; *Calendar of State Papers Colonial, America and West Indies,
1696–1697* (London, 1904), 617.
42. *Burlington Court Book*, 11, 19, 27; "Record of the Upland Court
. . ." Historical Society of Pennsylvania, *Memoirs*, VII (1860), 184,
194; *Chester County Court Records*, 14, 33, 69, 77; George, ed.,
Charter to William Penn, 136–37, 139, 140; PMHB, XXIII (1899),
403, 404; *Bucks County Court Records*, 10, 32, 43, 109–10; *Minutes of
the Provincial Council of Pennsylvania* (Philadelphia, 1852), I, 157.
43. Budd, *Good Order Established*, 9–10; Myers, ed., *Narratives of Pennsylvania*, 253, 290–91; PMHB, IV (1880), 198–99.
44. Carl Bridenbaugh, *Cities in the Wilderness: The First Century of
Urban Life in America, 1625–1742* (New York, 1938, 1966), 53; Bullivant, "Travel Diary," 69, 71; *Archives of Maryland*, XXIII (1903), 87;
XXV (1905), 386.
45. Carl Bridenbaugh, *Vexed and Troubled Englishmen, 1590–1642* (New
York, 1968), 472.
46. George Keith in Protestant Episcopal Historical Society, *Collections*,
I, xix, xx; and for the precedent set by the New England Yearly Meeting, see Carl Bridenbaugh, *Fat Mutton and Liberty of Conscience: Society in Rhode Island, 1636–1690* (Providence, 1974), 70–73.
47. Jones, *Quakers*, 438, 441, 522.
48. Budd, *Good Order Established*, 11–12.

III. RIGHT NEW-ENGLAND MEN

1. Smith's map (in its second state, 1614) is reproduced in William
Bradford, *History of Plymouth Plantation, 1620–1647*, ed. Worthing-
ton C. Ford (Boston, 1912), facing p. 189. It was published with the
explorer's *Description of New England* in which the name "Mas-
sachusets" (meaning at the Great Blue Hill of Milton) first appeared in
print. *Travels and Works of Captain John Smith*, ed. Edward Arber
and A. G. Bradley (Edinburgh, 1910), II, facing pp. 694, 699. In
homely phrase Captain Johnson wrote of "it being as unnatural for a
right N. E. man to live without an able Ministery, as for a Smith to
work his iron without a fire." *Johnson's Wonder-Working Providence,
1628–1651*, ed. J. Franklin Jameson, Original Narratives of Early
American History (New York, 1910), 214.
2. Of the islands in Massachusetts Bay, Captain John Smith wrote in *Advertisements for the unexperienced Planters of New-England, or any
where* (1631), "the land is of divers and sundry sorts, in some places
very blacke and fat, in others a good clay, sand and gravell . . . In the
Iles you may keepe your hogs, horse, cattell, conies or poultry, and
secure for little or nothing, and to command when you list; onely having a care of provision for some extraordinary cold winter." *Travels
and Works*, II, 949; [Samuel Maverick], "A Briefe Discription of New

England and the Several Towns therein, together with the present Government thereof," Massachusetts Historical Society, *Proceedings*, 2d Ser., I (1884–85), 237.

3. On the "Presettlement Forests of New England," consult Charles F. Carroll, *The Timber Economy of Puritan New England* (Providence, 1973), 25–37, 158–65.

4. Carl Bridenbaugh, *Vexed and Troubled Englishmen, 1590–1642* (New York, 1968), 395, 471–72; John G. Palfrey, *History of New England during the Stuart Dynasty* (Boston, 1859), II, 5–6; William G. Rossiter, *A Century of Population Growth* (Washington, 1909), pp. 4–5; *Johnson's Wonder-Working Providence*, 118.

5. Wallace M. Attwood, *The Physiographic Provinces of North America* (Boston, 1940), maps facing pp. 6, 8, 10, 12, and pp. 7, 11, 50, 67, 74, 99–100; Nevin M. Fenneman, *Physiography of the Eastern United States* (New York, 1938), plate 1, and pp. 370–78; Ralph Adams Brown, *Historical Geography of the United States* (New York, 1948), 7, 9, 13; John T. O'Neil in *New York Herald Tribune*, March 13, 1938, as cited by Jasper J. Stahl, *History of Old Broad Bay and Waldoboro* (Portland, Maine, 1956), I, 3.

6. For the spreading of settlement, see the series of maps, 1620–42, in Cambridge Historical Society, *Publications*, XXI (1930), 19–49; and maps for 1629, 1637, 1660, and 1700 in Lois K. Mathews, *The Expansion of New England* (New York, 1962), facing pp. 15, 23, 35, 66; Herman R. Friis, "A Series of Population Maps of the Colonies and the United States, 1625–1790," American Geographical Society, *Publication No. 3, Revised* (New York, 1968); William Hubbard, *A General History of New England* (Cambridge, 1815), 22; Colonial Society of Massachusetts, *Publications*, VI (1899–1900), 137–44.

7. *The Founding of Massachusetts*, ed. Stewart Mitchell (Boston, 1930), p. 45. From Great Wenham in Suffolk the vicar, James Hopkins wrote to Gov. John Winthrop, Feb. 25, 1632/33: "I hope you shall carrie a blessinge to your land, and the naturall inhabitants of it, which shall be your honour. If you can first civill the natives, and then bring some of them to know god . . . That plantation must needs prosper, where god (as I may say) hath an adventure." *Winthrop Papers* (Boston, 1943), III, 106; Joan E. M. Bellord, Puritan Ideas on Colonisation, 1620–1660 (M. A. thesis, University of London, 1951), 18–20, 23; Roger Williams to the General Court of Massachusetts Bay, Oct. 5, 1654, in "Letters of Roger Williams," ed. J. Russell Bartlett, *Narragansett Club Publications* (Providence, 1874), VI, 276.

8. Before the arrival of the English, the density of the Indian population of New England was probably "less than .22 persons per square mile." Alden T. Vaughan, *New England Frontier: Puritans and Indians, 1620–1675* (Boston, 1965), 28–29, 103.

9. For a map showing the locations of the principal New England tribes, see Vaughan, *New England Frontier*, 53. This work contains a more balanced discussion of both the Indians and the English (pp. 27–63)

than the recent shrill, bitter polemic of Francis Jennings, *The Invasion of America* (Chapel Hill, 1975).

10. Miss Bellord traces admirably the subtle shifting of views about relations with the Indians to 1660 in Puritan Ideas, 141–377. Her work is a notable contribution to Puritan studies.

11. Bellord, Puritan Ideas, 19, 38, 63, 68, 75, 195, 216; John Aylmer, (1558), quoted in Bridenbaugh, *Vexed and Troubled Englishmen*, 13.

12. The English scene from 1590 to 1642, in which these voluntary or forced exiles had been born and reared, is the subject of the writer's *Vexed and Troubled Englishmen;* for an elaboration of the conditions and events that led up to "The Puritan Hegira," see 434–76.

13. Bellord, Puritan Ideas, 262; Thomas Shepard, *Eye-Salve* (Boston, 1673), 12; William Stoughton, *New-Englands True Interest Not To Lie* (Boston, 1670), 15.

14. Sir Henry Vane to the Inhabitants of Rhode Island, Feb. 8, 1653/54, and the Town of Providence to Vane [prepared by Roger Williams at the request of the town] Aug. 27, 1654, "Letters of Roger Williams," 257, 267–68.

15. Contemporary English migrations to the Chesapeake region and the West Indies are treated in Bridenbaugh, *Vexed and Troubled Englishmen,* 394–443, and in Carl and Roberta Bridenbaugh, *No Peace Beyond the Line: The English in the Caribbean, 1624–1690* (New York, 1972).

16. Thomas Hutchinson, *The History of the Colony and Province of Massachusetts- Bay,* ed. Lawrence S. Mayo (Cambridge, 1936), I, 82; Herbert Moller, "Sex Composition and Correlated Culture Patterns of Colonial America," *William and Mary Quarterly,* 3d Ser., II (1945), 114–19.

17. *Johnson's Wonder-Working Providence,* 54; Hutchinson, *History,* I, 82n.

18. See John Winthrop, "A Modell of Christian Charity," (1630), in which the governor set down his Utopian concept of a heavenly society for New England, in *Winthrop Papers,* II (1932), 282–95; and Edward Johnson's more down-to-earth exposition of it for the common people in *Johnson's Wonder-Working Providence,* 23–43.

19. Cotton Mather, *Magnalia Christi Americana* (London, 1702), Book III, 118.

20. *Winthrop's Journal: "History of New England,"* ed. James K. Hosmer, Original Narratives of Early American History (New York, 1946), II, 31.

21. Carl Bridenbaugh, *The Spirit of '76: The Growth of American Patriotism Before Independence* (New York, 1975), 14–15; [Increase Mather], *The Necessity for Reformation, With Expedients subservient thereunto asserted* (Boston, 1679), Epistle Dedicatory.

IV. YANKEE USE AND ABUSE OF THE FOREST
IN THE BUILDING OF NEW ENGLAND, 1620–1660

1. *The Founding of Massachusetts: A Selection from the Sources of the History of the Settlement,* ed. Stewart Mitchell (Boston, 1930), 166.
2. *Records of the Governor and Company of the Massachusetts Bay in New England,* ed. Nathaniel B. Shurtleff (Boston, 1853–54), I, 28, 30. (Hereafter cited as: *Mass. Recs.*).
3. *Mass. Recs.,* I, 34, 42–44, 363–65.
4. *Founding of Massachusetts,* 166, 186, 187.
5. *Chronicles of the First Planters of the Colony of Massachusetts, Bay, from 1625 to 1636,* ed. Alexander Young (Boston, 1846), 378; *Johnson's Wonder-Working Providence, 1628–1651,* ed. J. Franklin Jameson, Original Narratives of Early American History (New York, 1910), 21, 22.
6. Christopher Levett, "A Voyage into New England," Massachusetts Historical Society, *Collections,* 3d Ser., VIII (1843), 180; John Josselyn, "Two Voyages to New-England," *ibid.,* III (1833), 295–96. For additional details, see three photographs of wigwams and early thatched-roof houses at the Colonial Village in Salem, in George F. Dow, *Every Day Life in the Massachusetts Bay Colony* (Boston, 1935), facing p. 16, and the superb photographs by Samuel Chamberlain in *Open House in New England* (Brattleboro, 1937), 37–88.
7. Cornelis van Tienhoven, Secretary of New Netherland, in an oft-quoted description of these cellar-houses in 1650, commented further: "The wealthy and principal men in New England, in the beginning of the Colonies, commenced their first dwelling-houses in this fashion for two reasons: first, in order not to waste time building and not to want food the next season; secondly, in order not to discourage poorer labouring people whom they brought over" with them. *Pennsylvania Archives,* 2d Ser. (Harrisburg, 1877), V, 182–83; *Johnson's Wonder-Working Providence,* pp. 111–12; John L. Sibley, *Biographical Sketches of Graduates of Harvard University* (Cambridge, 1873), I, 260; Norman M. Isham and Albert F. Brown, *Early Connecticut Houses* (Providence, 1900), fig. 2, p. 13; *Records of the Colony and Plantation of New Haven, from 1638–1649,* ed. Charles J. Hoadly (Hartford, 1857), I, 47, 70, 233.
8. *Mass. Recs.,* I, 363–65.
9. Until 1637, or soon thereafter, quantities of provisions to sustain the population were imported from Virginia, England, and Ireland, and for extended periods between sowing and harvesting, much of a family's time could be devoted to completing a house or barn, building fences, or clearing more land. *Johnson's Wonder-Working Providence,* pp. 22, 85, 118; Mass. Hist. Soc., *Colls.,* 4th Ser., IV (1858), 296; Richard Frothingham, *History of Charlestown* (Boston, 1845), p. 103; *Acts of the Privy Council: Colonial Series,* ed. W. L. Grant and James Munro (Hereford, 1908), I, 274, 275, 276, 278, 281–82, 283.

10. The figure for New Haven seems high, but De Vries was on the spot and much impressed by the many structures he saw. *Narratives of New Netherland, 1609–1664*, ed. J. Franklin Jameson, Original Narratives of Early American History (New York, 1909), 202, 205. For cottages at Plymouth and Salem, see Hugh Morrison, *Early American Architecture* (New York, 1952), figs. 4, 5, pp. 10–12, 30; James D. Phillips, *Salem in the Seventeenth Century* (Boston, 1933), 171–72.

11. The long-held misconception that the first New England settlers dwelt in log cabins was completely disproved in a notable study by Harold R. Shurtleff, *The Log-Cabin Myth* (Cambridge, 1939), 3–127, illustration facing p. 43; *Early Records of the Town of Dedham*(*1636*–*1706*), ed. D. G. Hill (Dedham, 1886–99), III, 29, 32, 33, 45.

12. *Suffolk Deeds*, ed. William B. Trask (Boston, 1880), I, 16; *Boston Records, 1634–1660 & Book of Possessions*, 2d Report of the Record Commissioners of the City of Boston (3d edn., Boston, 1902), 40; Mass. Hist. Soc., *Colls.*, 4th Ser., I (1852), 202.

13. *Tracts and other Papers Relating to the Origin, Settlement, and Progress of the Colonies in North America*, ed. Peter Force (New York, 1947), II, No. 4, p. 19; Essex Institute, *Historical Collections*, XLVII (1911), 2, LXXII (1946), 229–48; Morrison, *Early American Architecture*, figs. 12A, 23; *Winthrop's Journal: "History of New England": 1630–1649*, ed. James K. Hosmer, Original Narratives of Early American History (New York, 1957), I, 207.

14. William Wood, *New Englands Prospect* [1634], Prince Society, *Publications* (Boston, 1865), I, 53; *Winthrop Papers* (Boston, 1929–), III, 169; "Riving Shingles with a Frow," in H. R. Shurtleff, *Log-Cabin Myth*, facing p. 40.

15. Of the building of Charlestown, Edward Johnson wrote in 1648: "Poore Cottages them populate, with winters wet soone split, Brave Boston such beginning had, and Dorchester so began, Roxbury rose as mean as they." "Good News from New England," Mass. Hist. Soc., *Colls.*, 4th Ser., I, 201; *Winthrop Papers*, IV, 51, V, 291; Thomas Hutchinson, *History of the Colony and Province of Massachusetts-Bay*, ed. Lawrence S. Mayo (Cambridge, 1936), I, 48; "The Diaries of John Hull," American Antiquarian Society, *Archaeologia Americana*, III (1857), 169, 172, 199; William Hubbard, *General History of New England* (Cambridge, 1815), 199; *Winthrop's Journal*, I, 69–70, 155.

16. The housing and building conditions in England with which the Puritans were familiar before 1642 are discussed in Carl Bridenbaugh, *Vexed and Troubled Englishmen, 1590–1642* (New York and Oxford, 1968), 70–81.

17. *Mass. Recs.*, I, 74, 84, 109, 183; *Winthrop's Journal*, I, 112; II, 24.

18. In Connecticut, sawyers were not to be paid more than 6s. 6d. for "slit-work" (wood cut into thin deals or three-inch planks) for boards by the hundred. *Public Records of the Colony of Connecticut*, ed. J. H. Trumbull and C. J. Hoadly (Hartford, 1860–90), I, 65, 66; *Johnson's Wonder-Working Providence*, 199–200.

19. The crude construction of the first dwelling houses was repeated in nearly every new plantation, even by the Yankees of the third and fourth generations, as Madam Sarah Kemble Knight of Boston discovered on the Connecticut side of the Pawcatuck River in 1704: "This little Hutt was one of the wretchedest I ever saw a habitation for human creatures, It was supported with shores [corner posts] enclosed with Clapboards, laid on Lengthways, and so much asunder, that Light come throu' every where; the doors tyed on with a cord in the place of hinges; The floor the bear earth; no windows but such as the thin covering afforded." *The Journal of Madam Knight*, ed. Malcolm Freiberg (Boston, 1972), 13; *Johnson's Wonder-Working Providence*, 118 (italics mine).

20. *Conn. Recs.*, I, 8–9; Dow, *Every Day Life*, 18; "Note Book kept by Thomas Lechford, Esq.," in Amer. Antiq. Soc., *Archaeologia Americana*, VII (1885), 210.

21. "Lechford Note Book," VII, 59, 210.

22. Carl W. Condit, *American Building* (Chicago, 1968), 3–6, is an all too brief discussion of the shortage of proper tools before 1637. *Johnson's Wonder-Working Providence*, 232; "Good News from New England," I, 202, 209.

23. *Dedham Recs.*, III, 80–81, 106, 127; *New Haven Town Records, 1649–1662*, Ancient Town Records, I, ed. Franklin B. Dexter (New Haven, 1917), 151, 184, 224, 226, 269–70; Mason A Green, *Springfield, 1636–1886* (Springfield, 1888), 146; Dow, *Every Day Life*, 20.

24. *Mass. Recs.*, III, 181; Increase Mather, *Essay for the Recording of Illustrious Providences* (Boston, 1684), preface (n. pag.), 52–53, 59; *Early Records of the Town of Rowley, Massachusetts, 1639–1692*, ed. Benjamin P. Mighill and George P. Blodgett (Rowley, 1894), I, 91; "Hartford Town Votes," Connecticut Historical Society, *Collections*, VI (1897), 83; New-York Historical Society, *Collections* (1869), 235; *Winthrop's Journal*, I, 207.

25. New England Historical Genealogical Society, *Register*, V (1851), 443; *Provincial Papers: Documents and Records Relating to the Province of New Hampshire*, ed. Nathaniel Bouton (Concord, 1847–1941), XXXI (vol. I), 13; *Digest of Early Connecticut Probate Records*, ed. Charles W. Manwaring (Hartford, 1904), I, 15–16, 164.

26. *Winthrop Papers*, II, 516–17.

27. "Good News from New England," I, 204 (italics mine); Amer. Antiq. Soc., *Proceedings* (1943), LIII, 27, 221.

28. "Lechford Note Book," VII, 33, 252; *N. H. Provincial Papers*, XI, 75–76.

29. All temporary structures had disappeared from Salem before 1661. Essex County Deeds, Book V, leaf 107, cited by Dow, *Every Day Life*, p. 18; "Lechford Note Book," VII, 61, 63, 67; *Johnson's Wonder-Working Providence*, pp. 90, 96, 176, 211; *Conn. Probate Recs.*, I, 124; Hubbard, *General History*, p. 334.

30. *Winthrop Papers*, IV, map facing 416.

31. *Ibid.*, IV, 11–12.

32. "Lechford Note Book," VII, 302–3.

33. The Coddington window is preserved at the Rhode Island Historical Society. Howard M. Chapin, *Documentary History of Rhode Island* (Providence, 1919), II, plate facing p. 110; *Records of the Suffolk County Court, 1671–1680* Colonial Society of Massachusetts, *Publications*, XXX (1933), 861, 891; *Aspinwall Notarial Records from 1634 to 1651*, 32d Report of the Record Commissioners of the City of Boston (Boston, 1903), 40; *Mass. Recs.*, III, 102; Dow, *Every Day Life*, 21; *Ancient Historical Records of Norwalk, Conn.*, ed. Edwin Hall (Norwalk, 1847), 41–42; Mather, *Illustrious Providences*, 59.

34. To the people of this age, a *mansion* meant a separate country dwelling house or place of abode, though not a great stately residence or necessarily a *manor* house (*Oxford English Dictionary*). William Coddington, *A Demonstration of True Love* (London, 1674), 4; Antoinette Downing and Vincent Scully, *The Architectural Heritage of Newport, Rhode Island* (Cambridge, 1952), 22; plates 23, 24.

35. *The First Century of the History of Springfield. The Official Record from 1636 to 1736*, ed. Henry M. Burt (Springfield, 1898), I, 160–61.

36. Letter of John Eliot, Oct. 21, 1650, in Henry Whitfield, *The Light Appearing More and More Towards the Perfect Day* (London, 1651), 36–37; John Eliot, *A Farther Discovery of the Present State of the Indians in New England* (London, 1651), 36–37.

37. Mass. Hist. Soc., *Colls.*, 3d Ser., IV (1834), 177–78, 191; *Founding of Massachusetts*, 179.

38. George F. Dow, *Domestic Life in New England in the Seventeenth Century* (Topsfield, 1925), 2; John Pynchon, Account Book (Connecticut Valley Historical Museum, Springfield), II, 122, 123, 159, 223.

39. *Records of the Town of Jamaica, Long Island, New York, 1656–1751*, ed. Josephine C. Frost (Brooklyn, 1914), 15–16; *Southold Town Records*, ed. J. Wickham Case (Southold and Riverhead, 1882), I, 124–25; Frank C. Brown, "The Old House at Cutchogue, Long Island," *Old-Time New England*, XXXI (1940), 10–21; unidentified clipping from a New York newspaper of 1940.

40. No Renaissance classical elements appear in English vernacular building before 1660. On the International Vernacular, see Bridenbaugh, *Vexed and Troubled Englishmen*, 76–78. Downing and Scully, *Architectural Heritage of Newport*, 16–17.

41. Thomas F. Waters, "The Early Homes of the Puritans," Essex Inst., *Hist. Colls.*, XXXIII (1941), 52–53; *Records and Files of the Quarterly Courts of Essex County, Massachusetts*, ed. George F. Dow (Salem, 1911–21), II, 186.

42. Bridenbaugh, *Vexed and Troubled Englishmen*, 80–81.

43. The familiar rural New England sight of a farmhouse, sheds, and barn joined all in a row appears to have been a very late colonial development. Almost nothing is known about the arrangement of houses, barns, and outbuildings on the first farms, but at Reading, Mass. in

1649, the town meeting voted that "there being manni sad accidents in the Countree by fire, by joining of barnes and haystacks to dwelling howses, therfor no barne nor haystacke shall be set within six polles of anni dwelling howse opon penillte of twentie shillings." Lilly Eaton, *Genealogical History of the Town of Reading* (Boston, 1874), 8.

44. "Lechford Note Book," VII, 363; *Winthrop Papers*, IV, 418, 489; Rhode Island Land Evidences (transcript, Rhode Island Historical Society), I, 62; *Wyllys Papers*, Conn. Hist. Soc., *Colls.*, XXI (1924), 66, 124; Pynchon Account Book, I, 176; *Suffolk Deeds*, I, 130.

45. The town of Newbury voted to erect "a convenient schoolhouse" in 1652 but decided the next year to use the meetinghouse. Robert Paine built a schoolhouse for Ipswich and gave it to the Feoffees in 1652; a year later he presented a house for the schoolmaster with two acres of land. At Salem in 1655, the selectmen ordered materials and workmen "to repair *the town house* for the school and watch (my italics). The inhabitants of Rowley merely lent £5 to the schoolmaster, William Boynton, in 1657 "out of church stock" to build an addition to his own dwelling, provided he would agree to teach there for seven years. *Essex Quarterly Court Recs.*, I, 70, II, 232; N. E. Hist. Gen. Soc., *Reg.*, VI (1852), 65; "Town Records of Salem, 1634–1659," Essex Inst., *Colls.*, I (1868), 184; *Rowley Recs.*, I, 95; *Dorchester Town Records*, 4th Report of the Record Commissioners of the City of Boston (Boston, 1880), 54; William D. Love, *Colonial History of Hartford* (Hartford, 1914), 251–55; *Dedham Recs.*, III, 123, 136, 156–57, 182, 191; *Watertown Records* (Watertown, 1894), I, 18, 21; *Boston Recs.*, p. 109; *Book of Possessions*, 75.

46. George E. Littlefield, "Elijah Corlet and the 'Faire Grammar Schoole' at Cambridge," Col. Soc. Mass., *Pubs.*, XVII (1913–14), 133, 135–37; Lucius R. Paige, *History of Cambridge, Massachusetts* (Boston, 1877), 370–92.

47. John Pynchon of Springfield agreed in 1685 with Henry Stiles of Windsor "to dig my warehouse at the foot of the falls," 33′ x 18½′, "with the Passage or way into the Cellar 7½′ wide." One hundred and twenty loads of stone went into the structure at Warehouse Point on the Connecticut River. Pynchon, Account Book, VI, Pt. I, 282; "Lechford Note Book," VII, 69–71; *Boston Recs.*, pp. 63–64; *Book of Possessions*, pp. 9, 10, 21; *Suffolk Deeds*, I, 328, 335; *Winthrop Papers*, Mass. Hist. Soc., *Colls.*, 4th Ser., VII (1865), 231–32.

48. See illustrations and an account of the reconstruction of the structure in Percival H. Lombard, "The Aptuxcet Trading Post," *Old-Time New England*, XVIII (1917), 70–86; *Narratives of New Netherland*, 109–10; H. R. Shurtleff, *Log-Cabin Myth*, 68, 109, and illustration opposite; Morrison, *Early American Architecture*, 90; William Bradford, *Of Plymouth Plantation, 1620–1648*, ed. Samuel Eliot Morrison (New York, 1951), 259–79.

49. In 1635 John Friend built the fort at Saybrook, hitherto the largest building in New England, according to Samuel Eliot Morison, *The*

Founding of Harvard College (Cambridge, 1935), 222–23. That Friend was from Salem, not Ipswich, is clear from "Salem Recs" (I, 59, 71, 83, 175); in March 1640 he was admitted an inhabitant of Boston and also a member of the church at Salem. Apparently he was back on his farm at Salem in 1643, and he built several mills in Essex County before his death in 1656. N. E. Hist. Gen Soc., *Reg.*, I, 138, VI, 254; *Old-Time New England*, XXV, 117–18, XXXIII, 131–38; Morison, *Founding of Harvard College*, 271–77, 321, and illustrations of the Old College; *Harvard College Records*, Col. Soc. Mass., *Pubs.*, XIV (Boston, 1925), lxvi–lxvii, 17–18; *Records of the Colony of Plymouth in New England*, ed. Nathaniel B. Shurtleff (Boston, 1855–61), IX, 93–96; "Salem Recs.," I, 58, 71, 102, 175; *Johnson's Wonder-Working Providence*, p. 201; Cotton Mather, *Magnalia Christi Americana* (London, 1702), Bk. IV, 126.

50. *Boston Recs.*, 94, 134; Josiah H. Benton, *The Story of the Old Town House* (Boston, 1908), 49–55 (extracts from Keayne's will).

51. A reconstruction of the Boston Town House drawn from the original specifications is reproduced in Carl Bridenbaugh, *Cities in the Wilderness* (New York, 1939; Galaxy Paperback, 1966), facing p. 130. *Boston Recs.*, 134, 135; Benton, *Old Town House*, 58–67.

52. Mass. Hist. Soc., *Proceedings*, III (1855–58), 337–41; "Lechford Note Book," VII, 283–84; *Boston Recs.*, 71, 81, 84.

53. Noah Porter, *The New England Meeting House* (New Haven, 1933), 5; *Winthrop's Journal*, I, 75, 89.

54. Francis Johnson, *A Christian Plea Conteyning Three Treaties* (n. p. [Amsterdam?], 1617), 319.

55. Two excellent studies by Marian Card Donnelly bring together for the first time a body of evidence from which sound conclusions may be drawn: "New England Meetinghouses in the Seventeenth Century," *Old-Time New England* XLVII (1957), 85–97; and more fully in *The Meeting Houses of the Seventeenth Century* (Middletown, 1968), especially 5, 11, 12, 15. *Johnson's Wonder-Working Providence*, 73.

56. Mass. Hist. Soc., *Colls.*, 4th Ser., IV, 307.

57. John Pickering agreed with the town of Salem in Feb. 1638/39 to build an addition to the meetinghouse the breadth of the old structure of 1634 and 25' long to contain a gallery. There were to be "one Catted Chimney" 12' wide and to extend above the roof line, six windows, and a roof of 1½" planks. Pickering charged £63 in money to be received in three payments. "Salem Recs.," I, 81. Donnelly, *Meeting Houses*, 7, 12, 44; Donnelly, *Old-Time New England*, XLVII, 90, 97.

58. Some townsfolk preferred heat to light: and order was given at New Haven in Nov. 1651 that "the Casements of the Meeting House may have the glass taken out and boards fitted in, that in the winter it may bee warme; and in the summer they may bee taken downe to let in the ayre." *New Haven Recs.*, I, 98–99. Donnelly, *Meeting Houses*, 19; John G. Palfrey, *History of New England* (Boston, 1860), II, 59; Burt, *Springfield*, I, 176; *Records of the Town of Cambridge* (Cambridge, 1901), 85–86; *Watertown Recs.*, I, 38.

59. Captain Waldron was not the master builder at Dover as has been asserted (Sibley, *Harvard Graduates,* II [1881]. Job Lane's informative contract was first printed in the *Bi-Centennial Book of Malden* (Boston, 1850), 123–25. Mrs. Donnelly has a conjectural restoration of the Malden structure in *Old-Time New England,* XLVII, fig. 4, p. 91, XII, 27–32; Francis Baylies, *Historical Memoir of the Colony of New Plymouth* (Boston, 1830), 198; N.E. Hist. Gen. Soc., *Reg.,* XI (1857), 103–12; XIII (1859), 204.

60. See in general the important works by Charles F. Carroll, *The Timber Economy of Puritan New England* (Providence, 1973), and for richer detail, The Forest Society of New England: Timber, Trade, and Ecology in the Age of Wood, 1600–1691 (Ph.D. diss., Brown University, 1971). Ralph A. Brown, *Historical Geography of the United States* (New York, 1948), 108; For the fuel and timber shortages in England, see Bridenbaugh, *Vexed and Troubled Englishmen,* 64, 98–100, 149, 382.

61. Mills and farm buildings belonging to individuals who lived in a village near by were exempted from the half-mile rule. *Mass. Recs.,* I, 157, 292.

62. *Ancient Records of the Town of Ipswich,* ed. George Schofield (Ipswich, 1899), I, 9, 68, 98, 105; *Records of the Town of Plymouth* (Plymouth, 1889), I, 34; *Southold Recs.,* I, 319–20, 320–22; Hingham Town Records (Ms transcript, Mass. Hist., Soc.), 2, 110; *Dorchester Recs.,* 26, 91; *Winthrop Papers,* IV, 305; *N.H. Provincial Papers,* I, 138 (Exeter).

63. *Watertown Recs.,* I, 5, 22–23; *Mass. Recs.,* I, 144, 250; *Johnson's Wonder-Working Providence,* 113, 234.

64. *Mass. Recs.,* III, 158; *Watertown Recs.,* I, 14.

V. "THE FAMOUS INFAMOUS VAGRANT," TOM BELL

1. Paul Hazard, *European Thought in the Eighteenth Century from Montesquieu to Lessing* (New Haven, 1954), 249–54.

2. H. L. Beales in A. S. Turberville, ed., *Johnson's England* (Oxford, 1933), I, 143–46.

3. *An Apology for the Life of Mr. Bampfylde-Moore Carew, Commonly called the King of the Beggars* (London, 9th edn., printed for Robert Goadby, 1775), folded portrait frontispiece, and pp. 1–105.

4. Carew, *An Apology,* 106, 115–60.

5. Carew, *An Apology,* 161–77.

6. Carew, *An Apology,* 264–86; "Some Account of Bampfylde-Moore Carew," *Boston Weekly Post-Boy,* March 4, 1751; Charles Francis Adams, ed., *The Works of John Adams* (Boston, 1856), X, 219.

7. Robert Goadby brought out the first of many London editions in 1747. *Pennsylvania Gazette,* July 2, 1730; *Boston News-Letter,* April 19, August 2, 1733.

8. James Yates, an English schoolmaster, and Elizabeth Perry, both servants of Cornelius Van Horne of Upper Freehold, New Jersey, ran away in 1730. In an advertisement, the master offered £5 reward for their return and described them minutely. Yates is "a short round shoulder'd Fellow, stoops forward as he walks, of a dark Complexion and Pock-broken, speaks short and quick: He took with him a good Hat, two Wigs, one of natural Hair, the other a white one, a good mixt-colour'd homespun Drugget Coat with Buttons and Holes, old Shoes, and mixt blue and white Stockings. *He pretends to be a Scholar, and to have met with great Losses by his Travels and Trading.* He went in the Company with one Elizabeth Perry, Servant . . . an English Woman, aged about 20 Years, middle siz'd, fresh colour'd and squint ey'd: She had on a strip'd stuff Gown, a Cloak without a Cape, two strip'd Petticoats, a white Apron and a red and speckled one, strong low leather-heal'd Shoes." *Pennsylvania Gazette* [my italics], April 27, 1738; June 16, 1737; *Boston Evening Post,* November 29, 1736.

9. A preliminary report was read to the Massachusetts Historical Society on April 13, 1950, under the title "The Notorious Tom Bell—a Successful Harvard Man." The late Clifford K. Shipton's excellent sketch of Bell is in *Biographical Sketches of Those Who Attended Harvard College* (Cambridge, 1966), IX, 375–86; see also, Brooke E. Kleber on Bell in *Pennsylvania Magazine of History and Biography,* LXXV (1951), 416ff.

10. "Selectmen's Minutes, 1701–1715," in *Report of the Record Commissioners of the City of Boston* (Boston, 1884), XI, 108; "Selectmen's Minutes, 1716–1736," XIII (1885), 61; *William and Mary College Quarterly,* 2d ser., XXI (1921), 47; Suffolk Files (Court House, Boston), #29,931.

11. Shipton, *Harvard Graduates,* IX, 375; "Selectmen's Minutes, 1716–1736," 41, 131; "Selectmen's Minutes, 1701–1715," 61, 63, 120, 209; Kenneth B. Murdock "The Teaching of Latin and Greek at Boston Latin School," Colonial Society of Massachusetts, *Publications,* XXVII (1929–30), 22–25.

12. "Harvard College Records," Colonial Society of Massachusetts, *Publications,* XVII (1913–14), 275; Faculty Records (Harvard College Archives), I, 31, 32, 41.

13. Faculty Records, I, 46.

14. Steward's Ledger (Harvard College Archives), 46–47; Suffolk Files, #34,755; Shipton, *Harvard Graduates,* XI, 276.

15. *New-York Gazette and Weekly Post-Boy,* March 31, 1755.

16. *Virginia Gazette,* July 21, 1738.

17. Letter from New York in *Boston Gazette,* November 20, 1735.

18. *Pennsylvania Gazette,* May 31, June 14, 1739.

19. A petition, "signed by a considerable number of Merchants and other Inhabitants," was presented on August 30, 1739, to James Dottin, President of the Council of Barbados. It revealed a deep-seated anti-Semitism stemming from religious differences, economic competition,

and fear. One of the examples of Hebrew conduct adduced was: "The Jews confiding in their Numbers, do now daily attack, assault, and beat the Christians; whereas formerly, while less in Number they behav'd with Humility." Mention was made of "their daring Insolence to a stranger here," thereby introducing Tom Bell into the issue. Because England and Spain were at war in 1739, the presence of so many Jewish immigrants from the Spanish colonies, "in whom we can have no confidence," greatly alarmed the petitioners. In particular, they called for a law to prevent "aliens" from "dealing with Negroes." Two years later, on April 2, 1740, President Dottin notified the Board of Trade and the Duke of Newcastle that the Jews of Barbados were reported "to have almost entirely engrossed the business of Shop-keeping from the Christians, and . . . Subsist upon an unlawfull commerce with our Slaves, whom they encourage to commit Thefts and Robberies, and are the common receivers of all Stolen Goods." For the petition in full, see the *Virginia Gazette,* November 16, 1739; and for Dottin's letters: Public Record Office, London, C. O. 28: 25, Aa, 106; C. O. 28: 45, fol. 265, as quoted by Frank W. Pitman, *The Development of the British West Indies, 1700–1763* (New Haven, 1917), 30n. *Boston Evening Post,* September 10, 1739; promptly copied in Franklin's *Pennsylvania Gazette,* September 27, 1739.

20. Governor William Burnet's arms appeared on William Burgis's map of Boston, engraved by Thomas Johnston, 1728. (Copy at the Boston Athenaeum, which is reproduced by Walter M. Whitehill in *Boston: A Topographical History* (Cambridge, 1968), Fig. 17, p. 25. The Ledger of the Steward of Harvard College shows a credit "to Burgis" in Bell's final reckoning. He must have owned the print of 1728 and applied it against his debts. Shipton, *Harvard Graduates,* IX, 376; *Boston Evening Post,* September 10, December 10, 1739.

21. *Boston Evening Post,* December 10, 1739.

22. Report of Captain Rogers at Newport, March 7, 1739/40, as printed in the *Boston News-Letter,* March 13, and copied in the *New-York Gazette,* March 31, and the *Pennsylvania Gazette,* April 10, 1740. Lewis Timothy in the *South-Carolina Gazette* of September 12, 1743, erred in stating that Bell was released from standing in the pillory after the plea by the ladies; he was reprieved from being branded.

23. *New-York Weekly Journal,* February 15, 1742; *Pennsylvania Gazette,* February 24, 1741/2, February 12; 1742/3.

24. John Stockton, Esq., was the presiding judge of the Hunterdon Court of Common Pleas and the father of Richard Stockton, the Signer. The basic account of this notorious episode of mistaken identity was first published by Elias Boudinot as "Memoirs of the Rev. William Tennent," in the *Evangelical Intelligencer,* III (March 1806), 97–103. It should be read, however, in conjunction with Judge Richard S. Field's "Review of the Trial of the Rev. William Tennent for Perjury in 1742," New Jersey Historical Society, *Proceedings,* VI (1851–53), 31–40.

25. On the Rev. John Guild, see Shipton, *Harvard Graduates,* IX, 405–6;

Boudinot, "Memoirs," 191–92; Trenton Historical Society, *A History of Trenton, 1679–1929* (Princeton, 1929), II, 626–29.

26. Boudinot, "Memoirs," 192–193; Field, "Review of the Trial," 35–36.
27. Field, "Review of the Trial," 36–37; Boudinot, "Memoirs," 193.
28. In addition to the accounts by Boudinot and Field, the incomplete records at the New Jersey State Library, Trenton, contain the following: the indictment of William Tennent, Court of Common Pleas, Hunterdon County, August, 1741; Supreme Court Docket Book, #2, November, 1741; March, 1741–1742. Xerox copies of these were kindly provided for me by Dr. William C. Wright, Head, Bureau of Archives and History, State Library, for whose assistance I am grateful.
29. *Pennsylvania Gazette*, February 10, 1742/3; *New-York Weekly Journal*, February 15, 1742.
30. *New-York Weekly Journal*, February 15, 1742.
31. *Pennsylvania Journal*, February 8, 1742/3; *Pennsylvania Gazette*, February 10, 1742/3; *New-York Weekly Journal*, February 21, 1742/3; *Boston Weekly Post-Boy*, March 7, 1743; *Boston Weekly Magazine*, March 9, 1743; *South-Carolina Gazette*, April 18, 1743.
32. *Pennsylvania Gazette*, February 10, 1742/3.
33. *New-York Weekly Post-Boy*, February 12; also, *Boston News-Letter*, March 10, 1742/3.
34. News about Bell's escape from the prison at Philadelphia was slow in reaching Charleston, but in the *South-Carolina Gazette* for September 12, 1743, Lewis Timothy prefaced his account of June 16 with a warning that "As the notorious Tom Bell, may (in his present Travels) possibly make his Appearance here, we think the . . . Description of him from the *Philadelphia Gazette* [*sic*] will not be disagreeable, but perhaps useful to our Readers." *Pennsylvania Gazette*, June 16, 23, 30, July 14, 21, 1743; *Boston News-Letter*, July 7, 1743; *Pennsylvania Journal*, July 14, 21, 1743; *Boston Evening Post*, July 25, 1743.
35. *Boston Weekly Post-Boy*, August 22, 1743.
36. *Boston Evening Post*, September 12, October 3, 1743; *American Weekly Mercury*, September 22, 1743; *South-Carolina Gazette*, November 9, 1743; *Pennsylvania Journal*, October 3, 1743.
37. Middlesex County Court: General Sessions, 1744 (Court House, East Cambridge), under November 9, 10, December 9, 1743; King's Chapel, Boston: Register of Marriages, 1719–1841 (Massachusetts Historical Society).
38. *Boston Evening Post*, November 14, 1743; Middlesex General Sessions, 1744: November 9, 10, 18, 1743.
39. Middlesex General Sessions, 1744: December 9, 1743.
40. *Pennsylvania Gazette*, January 3, 1743/4; *Boston Evening Post*, December 12, 19, 26, 1743; *Pennsylvania Journal*, January 3, 1742/3; *New-York Weekly Post-Boy*, January 16, February 21, 1743/4.
41. *Boston Evening Post*, March 19, 1744; *New-York Weekly Post-Boy*,

April 2, 1744; *Pennsylvania Journal*, April 4, 1744; *Pennsylvania Gazette*, April 2, 1744; *American Weekly Mercury*, April 5, 1744.

42. The Petition of Thomas Bell, Middlesex General Sessions, 1744, probably to be dated March 15, 1743/4.

43. *Pennsylvania Gazette*, July 12, 1744; *New-York Weekly Post-Boy*, July 16, 1744; *Boston Evening Post*, July 23, 1744.

44. *Pennsylvania Gazette*, August 9, 1744; *Boston Evening Post*, August 13, 1744; *South-Carolina Gazette*, October 1, 1744.

45. *New-York Weekly Post-Boy*, September 17, November 5, 1744; *Boston News-Letter*, November 15, 1744.

46. Being "on the Pad" southward, Bell, posing in Maryland and Virginia as William Young, defrauded "several well-meaning Persons of considerable Sums." To preserve his reputation for integrity, Young advised the public through the *Maryland Gazette*, January 28, 1745, that "he never was in Virginia in his Life or in those parts of Maryland."

47. *South-Carolina Gazette*, February 11, 18, March 4, 11, 18, 1745. The news reached Northern readers promptly: *New-York Evening Post*, March 25, 1745; *New-York Weekly Post-Boy*, March 25, 1745; *Boston Evening Post*, March 25, 1745; *Boston Weekly Post-Boy*, April 8, 1745.

48. *Virginia Gazette*, October 31, November 7, 1745.

49. *New-York Weekly Post-Boy*, April 14, 1746; *Pennsylvania Journal*, April 17, 1746; *Virginia Gazette*, May 15, 1746.

50. *Pennsylvania Gazette*, August 14, September 18, 1746; *New-York Weekly Post-Boy*, September 15, 1746.

51. *New-York Weekly Post-Boy*, September 28, 1747; *Boston News-Letter*, October 8, November 19, 1747; *New-York Evening Post*, November 9, 1747; *Pennsylvania Gazette*, November 12, 1747.

52. [Boston] *Independent Advertiser*, October 3, 1748.

53. *Boston Evening Post*, January 30, 1749; *Pennsylvania Journal*, February 28, 1748/9. No copies of other Boston newspapers survive; and a search of the Suffolk and Middlesex county court files in March, 1979, failed to uncover any papers relating to the case. For two imitators of Bell's tactics at this time, one at Portsmouth, the other in Boston, see *Pennsylvania Journal*, August 24, 31, 1749.

54. *New-York Weekly Post-Boy*, August 28, 1749; *Pennsylvania Gazette*, August 31, 1749; *Boston Weekly Post-Boy*, September 4, 1749.

55. *New-York Evening Post*, September 4, 1749; also reprinted entire in the *Maryland Gazette*, September 27, 1749.

56. *New-York Evening Post*, September 4, 1749.

57. *New-York Evening Post*, September 4, 1749.

58. *New-York Weekly Post-Boy*, September 11, 1749; *Pennsylvania Gazette*, September 14, 1749.

59. It is just possible that Schoolmaster Bell managed to take his wife Esther to Hanover (or less likely that he took another mate there) to share in his return to respectability. At any rate, in October 1763, a

young man about twenty years of age, named Thomas Bell, a sadler born in Virginia, "five feet seven Inches high," stole a black horse during a journey in Georgia carrying a letter and £10.16.6 sterling from Mr. William Swinton of Sunbury to Mr. William Young of Savannah. The letter was found opened at the Savannah Ferry. If there is any validity in heredity, size, manner, and methods of cheating, we may suggest that this must have been Thomas Bell, Jr.—like father, like son. *Georgia Gazette*, October 12, November 17, 1763. Curiously, the *Boston Weekly Post-Boy* of March 4, 1752, contained a brief sketch of the exploits of Bampfylde-Moore Carew. *Virginia Gazette*, July 17, 1752; *Maryland Gazette*, August 13, 1752.

60. At this same time, women were also playing Tom Bell's game. Jonas Green of Annapolis printed an account of Charles Hamilton, a female masquerading as a man at Chester in Pennsylvania, claiming to be a physician. Actually, she had trained under a "noted Mountebank in England," been cast away off the Carolina Outer Banks, and worked her way northward through Virginia, Maryland, and the Lower Counties. When suspect and examined, she was found to be "a very bold" woman of twenty-three, though she appeared to be forty, named Charlotte Hamilton. *Maryland Gazette*, August 13, 1752; *Virginia Gazette*, August 28, 1752. At Penn's Manor, New Jersey, in October 1752, one Richard Perot, who "made so much Noise on pretence of being robb'd," turned out *"almost a second Tom Bell,"* according to the *New-York Weekly Post-Boy*, October 9, 1752 (my italics).

61. *Virginia Gazette*, April 14, 21, 28, 1752; *South-Carolina Gazette*, October 30, 1752; *Boston Weekly Post-Boy*, October 16, 1752.

62. *Virginia Gazette*, August 14, 21, 1752.

63. *Pensylvanische Berichte*, November 1, 1752.

64. *South-Carolina Gazette*, July 18, 1754.

65. *South-Carolina Gazette*, July 18, 25, August 1, 8, 1754.

66. *Antigua Gazette*, January 3, 1755, as clipped by the *New-York Weekly Post-Boy*, March 31, 1755.

67. *Pennsylvania Magazine of History & Biography*, III, 145. Copies of Mecom's "Dialogue" probably passed from hand-to-hand until they wore out. No copy has been found.

68. The fact that the printers of New York and Williamsburg, two of the principal scenes of Bell's rascalities, saw fit to print the letter strongly suggests that John Holt, Purdie, and Dixon believed the criminal to be their Tom Bell, who, when impersonation and cheating no longer paid off, reverted to his first calling of a mariner. *New-York Journal*, June 10, 1771; Purdie & Dixon, *Virginia Gazette*, July 4, 1771.

69. On intercolonial unity, see Carl Bridenbaugh, *The Spirit of '76: The Growth of American Patriotism before Independence, 1607–1776* (New York, 1975), 54–55, *et passim*.

VI. PHILOSOPHY PUT TO USE

1. "It seems, then, that nature has pointed out a mixed kind of life as
   most suitable to [the] human race, and secretly admonished them to
   allow none of these biasses to *draw* too much, so as to incapacitate
   them for other occupations and entertainments. Indulge your passions
   for science, says she, but let your science be human, and such as may
   have a direct reference to action and society. . . . Be a philosopher;
   but amidst all your philosophy be still a man." David Hume, "Inquiry
   Concerning Human Understanding," *Essays and Treatises Concerning Human Understanding* (Edinburgh, 1800), II, 7; I, 288–89,
   294–95; Leonard W. Labaree, ed., *The Papers of Benjamin Franklin*
   (New Haven, 1959–), IX, 251, italics mine; Jean François Marie
   Arouet de Voltaire, *Candide* (New York, 1929), 92–93.
2. *Pennsylvania Chronicle*, March 7, 1768.
3. On the issue of bishops for the colonies, see Carl Bridenbaugh, *Mitre
   and Sceptre: Transatlantic Faiths, Ideas, Personalities, and Politics,
   1690–1776* (New York, 1962, Galaxy paperback, 1967), 28–53, 83–115.
4. *Franklin Papers*, VI, 217; Leonard W. Labaree, ed., *The Autobiography of Benjamin Franklin* (New Haven, 1964), 105, 208.
5. *Franklin Autobiography*, 58n., 116–17; *Maryland Gazette*, March 24,
   1747; *A Collection of Several Pieces of Mr. John Locke Never before
   printed, or not extant in his Works* (London, 1720), 358–62.
6. *Franklin Autobiography*, 118; Carl and Jessica Bridenbaugh, *Rebels
   and Gentlemen: Philadelphia in the Age of Franklin* (New York, 1962,
   Galaxy paperback, 1962), x.
7. John Doddridge Humphreys, ed., *The Correspondence and Diary of
   Philip Doddridge* (London, 1829–31), II, 57; Paul Kaufman, "English
   Book Clubs and Their Role in Social History," *Libri*, XIV, 4–6, 21;
   Franklin Autobiography, 130.
8. This free translation of the difficult Latin of the motto is by Edwin
   Wolf 2d, *"At the Instance of Benjamin Franklin"*: A Brief History of
   the Library Company of Philadelphia, 1731–1976* (Philadelphia, 1976),
   2.
9. Stephen Bloore, "Joseph Breintnall, First Secretary of the Library
   Company," *The Pennsylvania Magazine of History and Biography*
   (PMHB), LIX (1924), 48; N. G. Brett-James, *Life of Peter Collinson*
   (London, 1925), 52, 62, 125; William Darlington, ed., *Memorials of
   John Bartram and Humphry Marshall* (Philadelphia, 1849), 124; "The
   First Books Imported by America's First Great Library, 1732,"
   PMHB, XXX (1906), 300–308; Wolf, *Library Company*, 3 (facsimile);
   *The Charter, Laws, and Catalogue of the Library Company of Phila-
   delphia* (Philadelphia, 1770).
10. Bloore, "Joseph Breintnall," 56; Darlington, *Memorials*, 60, 162; A
    Book of Minutes containing an Account of the Proceedings of the

Directors of the Library Company of Philadelphia, Beginning November 8, 1731, Library Company; Wolf, *Library Company*, 12–13.

11. Edwin Wolf 2d, *A Check-List of the Books in the Library Company of Philadelphia in and supplementary to Wing's Short-Title Catalogue, 1641–1700* (Philadelphia, 1959), iv; Carl Bridenbaugh, "The Press and the Book in Eighteenth Century Philadelphia," PMHB, LXV (1941), 17–24.

12. Jacob Duché, *Observations on a Variety of Subjects* (Philadelphia, 1774), II, 29–30.

13. Library Company, Minutes, I, 28.

14. Bloore, "Joseph Breintnall," 42–56; Library Company, Minutes, I, 27–28; Brett-James, *Peter Collinson*, 124; Darlington, *Memorials*, 61, 112n., 175, 316–17.

15. *Pennsylvania Gazette*, May 4, 1738; *Franklin Papers*, II, 2–7, 210. For two photographs of the cabinet as it now appears at the Library Company, see Wolf, *Library Company*, 6, 7, 54, where it is singled out as "the earliest example of American-made Palladian architectural furniture." Library Company, Minutes, I, 102; *Franklin Autobiography*, 117, 290.

16. *Franklin Papers*, II, 312; Library Company, Minutes, I, 114, 118–19, 193, 197; *Catalogue of Books of the Library Company of Philadelphia* (Philadelphia, 1764), 24–26; *Catalogue of the Library Company* (1770), 5.

17. *Pennsylvania Magazine, or American Monthly Museum*, I (February 1775), 53–57. For a careful study of "Franklin's Scientific Institution," that differs in emphasis and other respects from the present account, see Dorothy Grimm in *Pennsylvania History*, XXIII (1956), 437–62.

18. John Bartram to Peter Collinson, 1739, fragments in Bartram Papers, Historical Society of Pennsylvania (HSP), I, 38; III, 9, printed by Francis D. West in *Pennsylvania History*, XXIII (1956), 463–66; Darlington, *Memorials*, 132.

19. *Letters and Papers of Cadwallader Colden*, in New-York Historical Society, *Collections* (1918—), I, 272; II, 146, 208, 211, 247, 280; Darlington, *Memorials*, 142; Raymond P. Stearns, *Science in the British Colonies in America* (Urbana, 1970), 487; *Pennsylvania Gazette*, March 10, 17, 1742; *Pennsylvania History*, XXIII (1956), 464–66; *Franklin Papers*, II, 380–83.

20. John Bartram to Cadwallader Colden, April 7, 1745, Boston Public Library. Italics mine.

21. Peter Collinson had written to John Bartram in 1739 about the latter's proposal to found a philosophical academy in Philadelphia: "Your Library Company I take to be an essay towards such a Society." Darlington, *Memorials*, 132, 330; *Colden Papers*, III, 31, 34, 69, 143, 160, 330; Bartram to Colden, April 7, 1745, Boston Public Library.

22. All nine of the Philadelphia members of the ill-fated American Philosophical Society on April 5, 1744, also belonged to the Library Company. *Franklin Papers*, II, 406, 451n.; *Pennsylvania Gazette*, April 26,

May 3, July 26, 1744; Carl Bridenbaugh, ed., *Gentleman's Progress: The Itinerarium of Dr. Alexander Hamilton, 1744* (Chapel Hill, 1948), 189; *Franklin Autobiography*, 196; *Maryland Gazette*, Jan. 17, 1760; J. A. Leo Lemay, "Franklin's 'Dr. Spence,' " *Maryland Historical Magazine*, LIX (1959), 199–216.

23. "An Historical Account of the Wonderful Discoveries Made in Germany &C. Concerning Electricity," *The Gentleman's Magazine*, XV, 193–97; *Franklin Papers*, III, 118, 156; IV, 9; XVII, 65–66.

24. *Franklin Papers*, III, 118–19, 171, 364–65.

25. *Ibid.*, IV, 480n., 481; Bridenbaugh, *Rebels and Gentlemen*, 327–28, 331–32, 355: Shippen Papers, I, 131, HSP.

26. Library Company, Minutes, I, 145; Whitfield J. Bell, Jr., ed., *A Journey from Pennsylvania to Onandaga in 1743 by John Bartram, Lewis Evans, and Conrad Weiser* (Barre, Mass., 1973); Lawrence H. Gipson, *Lewis Evans* (Philadelphia, 1939), especially the five maps at the end of the volume.

27. *Pennsylvania Gazette*, Nov. 14, 1745; *Franklin Papers*, II, 419–46, and the diagrams of Lewis Evans facing pp. 432, 445; Adolph H. Benson, ed., *Peter Kalm's Travels in North America* (New York, 1964), I, 17, 25; II, 652–54; *Franklin Papers*, IV, 225–34; PMHB, XVII (1893), 273; [Benjamin Franklin], *Observations on the Increase of Mankind, Peopling Countries, &c.*, printed with William Clarke, *Observations on the Late and Present Conduct of the French . . .* (Boston, 1755; reprinted at London, 1755).

28. "Some Account of the Pennsylvania Hospital, 1754," *Franklin Papers*, V, 283–330; Thomas G. Morton and Frank Woodbury, *The History of the Pennsylvania Hospital* (rev. ed., Philadelphia, 1897), 1–36, and appendixes; [Benjamin Franklin], *Continuation of the Account of the Pennsylvania Hospital* (Philadelphia, 1761), 42, 44, 62, 66, 71–77; Bridenbaugh, *Rebels and Gentlemen*, 199–200, 244–47, 274–75, 290–91.

29. Du Simitière Scraps, 1752, No. 9, Library Company of Philadelphia, on deposit at HSP; *Maryland Gazette*, March 21, 1754; Nov. 4, 1773; *Pennsylvania Gazette*, Nov. 24, 1743; Bridenbaugh, *Rebels and Gentlemen*, 57–63.

30. For the period after June 1757, see in general, Bridenbaugh, *Rebels and Gentlemen*.

31. Abbé Gagliani to Mme. d'Épinay, from Naples, May 18, 1776, in *Abbé Ferdinand Galiani [sic] Correspondence avec Mme. d'Épinay . . .* , Lucien Perey and Gaston Maugras, eds. (Paris, 1889), II, 343; *Franklin Papers*, XVII, 126.

VII. THE NEW ENGLAND TOWN

1. For objectors, conscientious or otherwise, to the New England system, there remained Rhode Island, where township settlement lacked

the two basic characteristics of the rest of the area—fundamental
agreement of the body of the inhabitants in a religious and social
ideal, and identification of church and town. Since the Rhode Island
town represents, both historically and ideologically, an exception to
the New England way of life, I shall for the most part omit it from the
present discussion. It may also be mentioned that many towns in Connecticut consisted of a long street lined with houses rather than a
green as in New Haven.

2. In *Cities in Revolt: Urban Life in America, 1743–1776* (New York,
   1955, and Galaxy paperback, 1971).
3. Anon., *American Husbandry* (London, 1775), I, 61.
4. Stella M. Sutherland, *Population Distribution in Colonial America*
   (New York, 1937), xii; *Annual Report of the American Historical Association* (1931), 307.
5. Albert L. Olson, *Agricultural Economy and the Population in Eighteenth-Century Connecticut,* Connecticut Tercentenary Commission,
   No. 40 (New Haven, 1935), 18.
6. Ellen D. Larned, *History of Windham County, Connecticut* (Worcester, 1874), I, 263. There is a history of nearly every one of the 566
   New England towns of 1776. I have read a great many of them. In
   preparing this essay, however, I have leaned heavily on the works of
   Miss Larned because of her admirable quotations from the records.
7. *American Husbandry,* I, 62–65.
8. *American Husbandry,* I, 64; *Dictionary of American Biography,* XI,
   174–75; Frederick B. Goff, "Daniel Leonard et Alii" (Seminar Paper,
   Brown University, 1939); John Trumbull, *M'Fingal,* in V. L. Parring-
   ton, ed., *The Connecticut Wits* (New York, 1926), 66.
9. In 1756 Elizabeth Mosely sought in particular the custom of
   "Country-Gentlemen" at her Boston tavern. Charles Francis Adams,
   Three Episodes of Massachusetts History (Boston, 1892), II, 681–82;
   Boston News-Letter, Sept. 16, 1756.
10. *American Husbandry,* I, 66–67; Adams, *Three Episodes,* II, 685–86.
    Before the Revolution some rural families did not spend ten dollars a
    year. See "A Farmer" in *Hampshire Gazette,* Sept. 3, 1788.
11. This almanac is in the Rhode Island Historical Society.
12. "The trade of a Currier is very much wanted in Middletown, the Me-
    tropolis of Connecticut," where a master of that craft may "get a
    pretty Estate in a few Years," read an advertisement of 1758. *In-
    dependent Advertiser,* Mar. 13, 1749; *Boston Gazette,* Dec. 11, 1753;
    Jan. 14, 1760; July 29, 1765; William Davis Miller, "Samuel Casey,
    Silversmith," R. I. Hist. Soc., *Collections,* XXI, 1–14.
13. *American Husbandry,* I, 67.
14. Ann Hulton, *Letters of a Loyalist Lady* (Cambridge, 1927), 105.
15. Many years before, Roger Williams had said "Land is one of the Gods
    of New England." Quoted by Charles M. Andrews, *The Colonial
    Period of American History* (New Haven, 1934), I, 57n.

16. Large numbers of boys and girls, however, did hire out in their teens. *American Husbandry,* I, 70; Sutherland, *Population Distribution,* 47.
17. Larned, *Windham County,* I, 77; *Massachusetts Gazette and Post-Boy,* July 15, 1771.
18. *Boston Post-Boy,* Nov. 4, 1745.
19. Quoted by Samuel Eliot Morison in *Builders of the Bay Colony* (Boston, 1936), 62.
20. John Adams, *Works* (Boston, 1850–56), III, 400; John and Abigail Adams, *Familiar Letters During the Revolution* (New York, 1876), 120–21.
21. Almanac, in the Rhode Island Historical Society.
22. Adams, *Three Episodes,* III, 712.
23. "I am not a Man of Larning my selfe for I neaver had the advantage of six months schooling in my life. I am no travelor for I neaver was 50 Miles from where I was born in no direction, and I am no grate reader of antiant history for I always followed hard labour for a living. But I always thought it my duty to search into and see for my selfe in all maters that concansed me as a member of society, and when the war began between Britan and Amarica I was in the prime of Life and highly taken up with Liberty and a free Government. I see almost the first blood that was shed in Concord fite and scores of men dead, dying and wounded in the Cause of Libberty, which caused serious sencations in my mind." . . .
    "But I believed then and still believ it is a good cause which we ought to defend to the very last, and I have bin a Constant Reader of publick Newspapers and closely attended to men and measures ever since. . . ." William Manning, *The Key of Libberty,* ed. by S. E. Morison (Billerica, Mass., 1922), v–xiv, 3–4; S. E. Morison, "The Struggle over the adoption of the Constitution of Massachusetts, 1780," Mass. Hist. Soc., *Proceedings,* L, 354, 364, 401.
24. Hulton, *Letters of a Loyalist Lady,* 105; Noah Porter, *The New England Meeting House,* Connecticut Tercentenary Commission, No. 18 (New Haven, 1933), 5.
25. Thomas F. Waters, *Ipswich in the Massachusetts Bay Colony* (Ipswich, 1905), II, 3; Larned, *Windham County,* I, 60.
26. Quoted by Miss Larned, *Windham County,* I, 87.
27. Larned, *Windham County,* I, 281.
28. Alice M. Baldwin, *The New England Clergy and the American Revolution* (Durham, 1928).
29. Alexander Hamilton, *Itinerarium* (St. Louis, 1907), 148–49; Carl Bridenbaugh, ed., *Gentleman's Progress: The Itinerarium of Dr. Alexander Hamilton, 1744* (Chapel Hill, 1948), 121–22.
30. The social and political aspects of the militia system in the eighteenth century are worthy of study. For a brief sketch, see Allen French, *The First Year of the Revolution* (Boston, 1934), 32–36; also, Jonathan Smith, in Mass. Hist. Soc., *Proceedings,* LV, 345–46; *American*

Weekly Mercury, May 13, 1731; *Virginia Gazette*, Oct. 27, 1738. For a contrasting view, belittling the train bands, see C. F. Adams, *Three Episodes*, II, 264–65.

31. After 1754 advertisements appeared stating that "any Country town" destitute of a schoolmaster may have one "who will be faithful," on "Customary terms." Larned, *Windham County*, I, 262, 384, 550, 571, 572–73; Marcus W. Jernegan, *Laboring and Dependent Classes in Colonial America* (Chicago, 1931), 104, 121–26; Clifford K. Shipton, "Education in the Puritan Colonies," *New England Quarterly*, VII, 654–55; *Boston Gazette*, Dec. 17, 1754.

32. John Adams, *Works*, II, 85, 111–14, 125, 281; Timothy Dwight, *Travels in New England and New York* (New Haven, 1821–22), *passim*.

33. So great had become the traffic on the Massachusetts highways by 1765, that a caution was issued to all persons of "town and country" driving or walking always "to keep on the right-hand side of the way," as in England (which must have changed its rules at a later date!). *Boston Evening Post, Boston Gazette*, April 22, 1765. On Connecticut roads consult Isabel S. Mitchell's excellent *Roads and Road-Making in Colonial Connecticut*, Connecticut Tercentenary Commission, No. 14 (New Haven, 1933), especially the map. At Norwich Landing in Connecticut, on June 26, 1764, was raised Leffingwell's Bridge, "the most curious and compleat piece of architecture, of the kind, ever erected in America. It is 124 feet in length, 28 feet from the water [of the Thames River], and the water 30 feet deep; having nothing underneath between the buttments to support it, but is entirely supported by the geometry work above, and is supposed to be strong enough to bear 50 ton weight. The work was performed by Mr. John Bliss of Norwich, said to be one of the most curious mechanicks this age has produced." *Boston News-Letter*, July 19, 1764.

34. Improved roads and the stage coach, for example, made possible the rise of Stafford and Mansfield Springs, in Connecticut, as summer resorts. See the author's "Colonial Newport as a Summer Resort," *Rhode Island Historical Society Collections*, XXVI (January 1933), 1–23.

35. On the post riders, see *Boston Post-Boy*, Dec. 25, 1749; *Boston Gazette*, July 27, 1767; and Hugh Finlay, *Journal* (Brooklyn, 1867). For the Boston-Providence-Newport stage, consult *Boston News-Letter*, Aug. 8, 1745; May 26, 1757; *Weekly Advertiser*, July 17, 1758; *Boston Post-Boy*, July 20, 1767; *Boston Gazette*, Aug., 17, 1767. The stage to Portsmouth, N. H., began to operate in 1763. *Boston News-Letter*, Apr. 14, 1763. Silent Wilde was the regular "carrier" from Boston to Deerfield, 1761–75. *Boston Post-Boy*, April 6, 1761; *Boston Chronicle*, Oct. 26, 1775.

36. The Amorys opened a branch store at Salem. At Charlestown Ferry, Charles Russell, apothecary, made a special point of outfitting "Gentlemen Practitioners . . . in the Country." *Boston News-Letter*, Nov.

14, 1754; Oct. 12, 1758; Jan. 15, Nov. 26, 1751; *Boston Post-Boy*, Dec. 26, 1774; *Boston Gazette*, May 7, 1764.

37. Moses Holmes of Ashford covered the route from Boston to Hartford; Christopher Page to Springfield, Samuel Farrar to Fitchburg, and Silent Wilde to Deerfield. Wilde also chaperoned Deerfield young ladies on trips to and from Boston! *Weekly Advertiser*, Aug. 14, 1758; Nov. 12, 1759; *Boston News-Letter*, Dec. 31, 1761; July 3, 1766; *Boston Post-Boy*, Feb. 18, May 27, 1771; Col. Soc. Mass., *Publications*, XI, 34, 34n.

38. Bridenbaugh, *Cities in the Wilderness*, 452; *Boston Post-Boy*, May 3, 1762; April 18, 1767; *Boston News-Letter*, May 13, 1762.

39. A Catalogue of Mein's Circulating Library (Ms. in Mass. Hist. Soc.); *Boston Gazette*, July 13, 1761; *Boston Post-Boy*, Feb. 1, 1762; Nov. 4, 11, 1765; May 19, 1766; *Boston News-Letter*, Oct. 31, 1765; March 3, 31, 1774. For social libraries in rural towns, see Larned, *Windham County*, I, 356–59, 526, 574–77. Darien, Connecticut, seems to have had a "Book Company" as early as 1733. Joseph T. Wheeler, Literary Culture in Colonial Maryland (Ms Thesis, Brown University, 1938), 5.

40. Mr. Dwight's quaint conceits annoyed the ministerial association, which visited him and rebuked him. The parson meekly admitted his fault and promised reform, but in praying at the departure of the worthies "hoped that they might so hitch their horses together on earth that they should never kick in the stables of everlasting salvation." Quoted by Larned, *Windham County*, I, 51.

41. For the standard traveler's tale about Yankee curiosity, see Patrick M'Robert, *A Tour through Part of the North Provinces of America . . . 1774 & 1775* (Philadelphia, 1935; New York, 1968), 15–16.

42. Windham, a shire-town of Connecticut, had voted in 1728: "That all baptismal persons have a right to hear confessions for public scandal, and that no such confessions shall be accepted unless made before the congregation on the Sabbath, or some public meeting wherein all baptized persons have a warning to attend." Public penance descended from medieval Roman Catholic practice via the Church of England to the Puritans in New England. Quotation in Larned, *Windham County*, I, 271; see also, Adams, *Three Episodes*, II, 795; Mass. Hist. Soc., *Proceedings*, II, 477–516.

43. Larned, *Windham County*, I, 273.

44. *Massachusetts Gazette and Post-Boy*, Nov. 18, 1771.

45. The *Boston News-Letter*, in 1750, published "Rules to be observed in TRADE," urging its readers to "Strive to maintain a *fair character* in the world." *Boston News-Letter*, July 12, 1744; Mar. 29, 1750; Oct. 11, Nov. 8, 1759; June 20, 1765; *New-York Evening Post*, Nov. 14, 1748; *Newport Mercury*, Dec. 24, 1764.

46. C. K. Shipton, *Sibley's Harvard Graduates* (Cambridge, 1933–1945), IV, 4, 251; VII, 202; Larned, *Windham County*, I, 60–61.

47. Quoted in Adams, *Three Episodes*, II, 618.

VIII. VIOLENCE AND VIRTUE IN VIRGINIA, 1766

1. Through his mother, born Mary Sowerby, Robert Routledge was own cousin to James Wallace, attorney general to the British government, 1780, 1783; and Wallace's son was created Baron Wallace of Knaresdale in 1828. Another cousin, Anne Plasket, married John Law, D.D., Bishop of Elphin and brother of Lord Ellenborough, Lord Chief Justice of the King's Bench. Amid such respectability, illegitimacy proved no problem in those days, for Robert's son John prospered as a yeoman farmer in the parish of Bewcastle in Cumberland. John V. Harrison (whose wife is a descendant of John Routledge) to Secretary of the Virginia Historical Society, October 20, 1951; January 26, April 22, 1952; Mr. Harrison to the writer, November 9, 1963. I am much indebted to our fellow Member John Melville Jennings for assistance in this and other connections.

2. Samuel Shepard, *Virginia Statutes at Large* (Richmond, 1792), I, 315–16; "Dikephilos," in Purdie & Dixon's *Virginia Gazette*, July 18, August 29, 1766; Herbert C. Bradshaw, *History of Prince Edward County, Virginia* (Richmond, 1955), 100–102.

3. Chiswell was apparently pronounced "Chizzel," for so George Washington spelled it phonetically in his diary.

4. The account given above is a paraphrase of the reliable report of the affair given by "Dikephilos," which is adjusted in a few minor details by other descriptions also in P. & D. *Va. Gaz.*, June 20, July 4, 18, October 10, 1766.

5. "Dikephilos" (a lover of justice) is discovered to have been the Reverend Jonathan Boucher in *Reminiscences of an American Loyalist, 1758–1789*, ed., Jonathan Bouchier (Boston, 1925), 111.

6. Colonel Chiswell's counsel, John Wayles, challenged the accuracy of "Dikephilos" at this point by insisting "That upon something Chiswell said (as calling him a fellow) the deceased turned suddenly around, and then Chiswell dropped the point of his sword, and held it with an extended arm without advancing. That when Routledge sunk in Carrington's arms, he said he believed the man was dead; yes, replied Chiswell, you may take him away, I have killed him. The reason of saying which was, Chiswell said he felt him on the point of his sword, which no other man could know." In other words, Routledge ran onto the sword, and this was the defense Chiswell's friends planned to make for him. Wayles's version was demolished by other writers, and the irate counsel sued certain parties for libel, as will appear. P. & D. *Va. Gaz.*, July 18, 1766, p. 2, for the murder diagram; also see, Sept. 12, Oct. 10, 1766.

7. The eight witnesses were each bound in £50 to appear at Williamsburg on the sixth day of the next General Court to give evidence in this case. The Examining Court was made up of Piedmont planters: John Fleming, John Netherland, Thomas Tabb, Carter Henry Harrison, John Mayo, William Smith, and John Woodson, "Gentlemen."

P. & D. *Va. Gaz.*, October 10, 1766; Cumberland County, Order Book, 1764–1767 (Virginia State Library), 253, 258. I am obligated to Mr. William J. Van Schreven for bringing this material to my attention.

8. The judges did talk with Sheriff Thomas and John Wayles, attorney for John Chiswell, both of whom had been present at the Examining Court but not at the killing of Routledge. P. & D. *Va. Gaz.*, June 20, July 4, 18, 1766.

9. The sources and the nature of the tensions agitating various groups have been discussed by the writer in Chapter 6 of *Seat of Empire: The Political Role of Williamsburg*, 2d edn., rev. (Williamsburg, 1958); and from a somewhat different point of view in *Myths and Realities: Societies of the Colonial South* (Baton Rouge, 1952), Chapter I: "The Chesapeake Society."

10. For the Stamp Act and Virginia, see *Journals of the House of Burgesses of Virginia, 1761–1765*, ed. John Pendleton Kennedy (Richmond, 1907); Edmund S. and Helen M. Morgan, *The Stamp Act Crisis* (Chapel Hill, 1953), 96–97; Bridenbaugh, *Seat of Empire*, 55–62; Edmund Randolph, History of Virginia (Virginia Historical Society), 104–8.

11. Rind's *Virginia Gazette*, May 16, 1766; P. & D. *Va. Gaz.*, May 16, 1766; *Journals of the House of Burgesses of Virginia, 1766–1769*, ed. John Pendleton Kennedy (Richmond, 1906), x–xiii, xxvi; Randolph, History of Virginia, 110–11; and for the standard account of the Robinson affair, David J. Mays, *Edmund Pendleton, 1721–1803, A Biography* (Cambridge, 1952), I, 174–223, and Appendices II–VI.

12. *Virginia Magazine of History and Biography*, IV (1896–97), 359; *Executive Journals of the Council of Colonial Virginia*, ed. H. R. McIlwaine, IV (Richmond, 1930), 27, 83, 128, 179, 258, 258, 350, 380.

13. *Exec. Journals*, IV, 391, 396; *Va. Gaz.*, Dec. 14, 1739; William G. Stannard, *Colonial Virginia Register* (Albany, 1902), 114–46.

14. *William and Mary College Quarterly*, II (1893–94), 235–36; XIV (1905–6), 213; Rind, *Va. Gaz.*, June 2, 1768; P. & D. *Va. Gaz.*, November 29, 1770. For "Scotchtown" and the Chiswell house in Williamsburg, see Thomas T. Waterman, *The Mansions of Virginia* (Chapel Hill, 1945), 65–66, 74, 85, 415, 422; and *Colonial Williamsburg: Official Guide Book & Map* (Williamsburg, 1957), 63, 64, and map, for the reconstructed house.

15. Reverend Robert Rose, Diary (Typescript: Colonial Williamsburg Research Library), July 31, 1750, p. 163.

16. *Wm. & Mary College Quarterly*, XIV, 142–43; *The Burd Papers: Extracts from Chief Justice William Allen's Letter Book*, ed. Lewis B. Walker (Pottsville, Pa., 1897), 16; and for the intercolonial money transactions, see Carl Bridenbaugh, *Cities in Revolt: Urban Life in America, 1743–1776* (New York, 1958; Galaxy paperback, 1971), 94.

17. Pleadings of Robinson Adminrs vs Byrd et als., P. B. Carrington Letter Book (Virginia Historical Society), 209–11, 219–20; *Va. Gaz.*, Sep-

tember 2, 1757; William W. Hening, *The Statutes at Large of Virginia* (Philadelphia, 1823), VIII, 270.

18. Chiswell's Mine was an unusually large venture for a frontier community. It used some thirty-six slaves, two wagons, and sixteen horses. Carrington, Letter Book, 211–14, 224; and for the debt to the Robinson estate, see Mays, *Edmund Pendleton*, I, 203, 364.

19. The Assembly passed an act within the month to vest in trustees the lands of "John Chiswell, Gentleman," of Williamsburg. Hening, *Statutes*, VIII, 270; Will of John Chiswell, Land Office, Miscellaneous Wills, 63–888 (Virginia State Library).

20. *William and Mary Quarterly*, XXI (1912–13), 164; Clarence S. Brigham, *History and Bibliography of American Newspapers, 1690–1820* (Worcester, 1947), II, 1159.

21. *Va. Gaz.*, March 7, 1766; Rind *Va. Gaz.*, No. 1, May 16, 1766.

22. *Newport Mercury*, January 26, February 2, 1766.

23. Two weeks after the story broke, Councilor and Judge John Blair sent in a reply, which he had prepared because the granting of bail by him and his two colleagues had been so "much censured by many people" and severely condemned by an anonymous correspondent in his "Warm zeal" against their act. He based his defense upon an inaccurate account of the murder from which the judges had concluded that it was "an unhappy drunken affair" and legally bailable. This communication was reported in full by the *Maryland Gazette*, July 17, 1766; P. & D. *Va. Gaz.*, June 20, July 4, 1766.

24. Unfortunately, only ten issues of William Rind's newspaper have come down to us. They are: May 16, 30, July 18, August 8, 15 (Supplement), September 5, November 27, December 4, 11 and Supplement, 25, 1766. It is clear, however, that Purdie & Dixon printed most of the controversial pieces.

25. P. & D. *Va. Gaz.*, July 11, 1766.

26. In preparing his opinion, George Wythe evidently overlooked the right of appeal from decisions of colonial courts to the Privy Council. P. & D. *Va. Gaz.*, August 1, 1766.

27. "A Friend to the Constitution" replied to "Manners" with the argument that the central issue was whether the Examining Court was a court of record. If so, then it could be reversed only by a judgment of the General Court, "Who have the Supreme controuling power over all the County Courts in the colony. . . . It is useless to ransack books of law for cases relative to this question, for unless you can show that courts like our County Courts are established in England, and that their judgments have been countrouled by the Judges of the King's Bench at their chambers, all your learning will, I presume, be to little purpose." He concluded by saying that he did not impeach the judgments of any persons, or desire to enter into "party disputes" (which he frankly recognized as raging in the Old Dominion). P. & D. *Va. Gaz.*, August 29, September 12, October 30, 1766.

28. "This motley kind of court called the General Court," Quincy noted,

"is composed of the Governor and Council, who are appointed and created by mandamus from the Crown, and hold *bene placito*. I am told it is no uncommon thing for this Court to sit one hour and hear a cause as a Court of Law; and the next hour, perhaps minute, to sit and audit the same matter as a Court of Chancery and equity: and if my information is good, they very frequently give directly contrary decisions. . . . An aristocratic spirit and principle is very prevalent in the laws, polity, and manners of this Colony." Josiah Quincy, Jr., "Journal," Massachusetts Historical Society, *Proceedings*, XLIX (1915–16), 465–66; P. & D. *Va. Gaz.*, September 12, 1766.

29. P. & D. *Va. Gaz.*, July 11, 18, August 29, 1766.
30. P. & D. *Va. Gaz.*, July 18, 1766.
31. *William and Mary College Quarterly*, II, 239; Robert Carter to Thomas Bladen, Esqr., Robert Carter, Letter Book, 1764–1768 (Colonial Williamsburg Research Library); Stannard, *Colonial Virginia Register*, 49.
32. My italics. P & D., *Va. Gaz.*, July 25, 1766.
33. "The King's Attorney going out of the country the very morning of the day when the said [Examining] Court was held and leaving no one to officiate for him, must, in the eyes of all men who are not blind, look extremely dark." P. & D. *Va. Gaz.*, September 19, 1766. For other communications on this subject in the same paper, see September 5, 12, October 10, 17, 1766.
34. Quotation from Bridenbaugh, *Seat of Empire*, 69. For newspaper exchanges in the Northern Colonies, see the writer's *Mitre and Sceptre: Transatlantic Faiths, Ideas, Personalities, and Politics, 1689–1775* (New York, 1962), especially Chapters VI, VIII, XI.
35. In all probability, the author of "From the East" and "A Prophecy from the East" was the Rev. John Camm. P. & D. *Va. Gaz.*, and Rind *Va. Gaz.*, August 15, Supplements, 1766.
36. P. & D., *Va. Gaz.*, September 12, 1766.
37. P. & D. *Va. Gaz.*, October 17, 1766; also *Maryland Gazette* (from Rind), October 30, 1766.
38. In 1850, *eighty-four years* after Chiswell's death, the Reverend John B. Dabney wrote Sketches and Reminiscences of the Dabney and Morris Families, in which he gave such an inaccurate and garbled account of the murder of Robert Routledge as to bring into question his report of the death and burial of John Chiswell. But because this dubious "source" has previously been used uncritically to prove suicide, it is given below for what it may be worth:

"His remains were brought up for interment to Scotch-town, his seat in the County of Hanover. A report got abroad that the alledged suicide was a mere strategem to facilitate the escape of the murderer from the just punishment of his crime. An immense crowd assembled on the occasion, and the multitude insisted that the coffin should be opened, believing that it was either empty, or occupied by some substitute for the body of the supposed self-destroyer. Prepossessed with this idea, the spectators could not be persuaded that, in the black and distorted features of the corse [sic] before them, they beheld the genuine relics of that proud man, whose execution they had so lately demanded as an expiation due to the offended majesty of justice. When such a suspicion, however groundless, has once entered the minds of a mob, they cling to it with a tenacity proportioned to the vehemence of their excitement; and the greater its improbability, the more implicit is their belief. . . . The controversy ran so high, that the ceremonies of interment were suspended, and the people declared they would not permit the body to be buried until it was identified by Col. William Dabney. It was known that he was a relative of the deceased, and on terms of intimate acquaintance with him; yet there was not one in that exasperated assemblage, who had not the firmest reliance on his truth and candor. . . . After inspecting the corpse at the request of the bystanders, he pronounced it to be the body of Col. Chiswell, and the temper of the popular commotion was instantly quelled. These particulars I have derived from several members of our family, and I have no doubt of their authenticity." Pp. 6–7 of the copy in Colonial Williamsburg Research Library, taken from the original MS lent by Mrs. Albert M. Pennybacker of 3224 Winsor Court, Chattanooga, Tenn. See also, Boucher, *Reminiscences of a Loyalist*, 111, for the earliest known expression of the doubt quoted above.

39. P. & D. *Va. Gaz.*, October 17, 1766; *Maryland Gazette*, October 30, 1766. For the pieces of Wayles and Blair, see P. & D. *Va. Gaz.*, September 12, 19, October 10, 1766.

40. At this time, in a letter to a kinsman, William Nelson privately expressed similar views, albeit with more restraint than "R. R.": "I see you have had a view of our papers which would let you into our political Disputes. *They are numerous* and too scurrilous to merit much of your attention." To John Nelson, November 12, 1766, William Nelson, Letter Book (Virginia State Library), 16 (my italics); P. & D. *Va. Gaz.*, October 10, November 6, December 18, 1766.

41. William Rind, his widow Clementina, and successors continued their *Virginia Gazette* until 1776. *Journals of the House of Burgesses of Virginia*, 1766–1769, 17–18; P. & D. *Va. Gaz.*, November 27, 1766; Brigham, *Bibliography*, II, 1145, 1159–63.

42. P. & D. *Va. Gaz.*, November 6, December 18, 1766.

43. By means of a paid notice, Elizabeth Barebones complimented William Rind upon getting the public printing, but wanted to know what happened to a communication sent to him long since and hoped that "there are no public licenses behind the curtain." P. & D. *Va. Gaz.*, August 22, November 6, 1766; November 20, 1768.

44. Smarting under Jonathan Boucher's taunt about leaving the country, "R. R." attacked "Dikephilos" in verses over the pseudonym "A Man of Principle" in Rind's *Virginia Gazette*, and "A Foe to Pedants" (possibly also "R. R.") had this to say, which elicited the reply of "Marcus

Curtius" mentioned above: "How long, Dikephilos, must we sustain/ The venal efforts of thy flimsy Brain!" Rind, *Va. Gaz.*, November 27, December 25, 1766; P. & D. *Va. Gaz.*, January 1, 1767; David Hume, *Essays and Treatises on Several Subjects* (Edinburgh, 1800), I, 9–12.

45. Robert Routledge had died "intestate and without heirs"; therefore his 1,272 acres of land escheated to the colony and in 1776 passed on to the Commonwealth of Virginia. They were put to good use in 1794, when the Assembly passed an act granting them to Hampden Sydney College to "promote the education of youth." Shepherd, *Statutes at Large*, I, 315–16.

46. Although all attempts at secrecy failed at the time, Virginia historians and gentry have studiously avoided any mention of this aristocratic murder ever since. As recently as 1949 the writer was told by a Virginian that this was not the kind of a matter to be brought out into the open: "It is not history." Not until 1955 did Bradshaw supply some details in his *History of Prince Edward County, Virginia.* Curiously, one of the central characters in the Routledge-Chiswell affair was the victim of the next aristocratic homicide in Virginia to be completely hushed up: George Wythe. The crime, which occurred on May 25, 1806, is fascinatingly uncovered by Julian P. Boyd in *The Murder of George Wythe* (Philadelphia, 1949).

IX. BATHS AND WATERING PLACES OF COLONIAL AMERICA

1. For early springs, see Carl Bridenbaugh, *Cities in the Wilderness: The First Century of Urban Life in America, 1625–1742* (New York, 1938), 118, 276, 439–40; and *Maryland Archives*, XXII, 279; XXV, 17, 31.

2. On the rage for the "Cold Bath", see [Boston] *Independent Advertiser*, May 30, 1748; *Boston Post-Boy*, June 16, 1766.

3. Jonathan Trumbull, *Complete History of Connecticut* (2 vols., New London, 1898), II, 60.

4. Samuel Peters, *A General History of Connecticut* (London, 1782), 174.

5. "These waters are very improper for young people of a full habit of body . . . ," and "are not to be wantonly used or in sport." Benjamin Mecom of the *Connecticut Gazette* offered to print cases of cures sent to Mr. Guy. *Conn. Gazette*, July 26, Aug. 9, 1766; *Boston Gazette*, Aug. 4, 1766; *Newport Mercury*, Aug. 11, 1766.

6. *Bos. Gazette*, June 1, 1767.

7. In the same issue as Jackson's advertisement appeared the following from a leading physician: "Whereas a Report has prevailed that I have said that Mr. Jackson's Mineral Springs at New-Boston arose from Copper, and was of a noxious Quality: I therefore take this Method to inform the Public, that I have found by Experiment that those Waters

are strongly impregnated with Steel, and that I have recommended the use of them to such of my patients as I thought required Chalybeate Medicines. And am of the Opinion that they may be serviceable in many Disorders, if properly used.—James Lloyd." Dr. William Lee Perkins also endorsed the spring. *Boston News-Letter*, Aug. 6, 1767; *Bos. Gazette*, Aug. 17, 31, 1767.

8. *Bos. Gazette*, Aug. 4, 11, 1766; *Newport Mercury*, Aug. 16, 1767; *Conn. Gazette*, Aug. 22, 1767. Besides misuse there was fraud. Some Boston invalids, unable to go to Stafford, hired a carter to procure the mineral water. On his way home many people importuned him to sell them some, which he did. At every brook this slick Yankee refilled the barrel, so that he boasted a turnover of 160 gallons. His Boston clients, never suspecting, drank copiously of the water he delivered and declared themselves benefited! *Bos. Gazette*, Aug. 4, 1766.

9. *Bos. Post-Boy*, Sept. 6, 1766.

10. *Bos. News-Letter*, July 2, 1767.

11. *Bos. News-Letter*, July 9, 1767.

12. *Conn. Gazette*, Aug. 9, 1766; *Bos. Gazette*, Aug. 31, 1767; *Bos. News-Letter*, Sept. 3, 1767.

13. This was travel *de luxe*. The baggage wagon departed sixteen days ahead of the coach. From January 17 on, Wood also ran two round trips daily to Newton Springs. *Bos. News-Letter*, Apr. 23, 1767; *Bos. Gazette*, May 4, June 8, 1767.

14. Two weeks later the *Chronicle* reported a death from bathing. *Bos. Chronicle*, Aug. 15, 29, 1768; *Bos. Evening Post*, June, 1768.

15. Via Worcester, Spencer, Brookfield, Brimfield, and Western.

16. John Adams, *Works* (10 vols., Boston, 1850–56), II, 267.

17. Adams, *Works*, II, 267.

18. Adams, *Works*, II, 288.

19. Adams relates several excellent anecdotes about persons he met on this trip. *Works*, II, 264n., 268.

20. The Rev. Mr. Peters, in his *General History* (240), says that Stafford was the "hospital of the invalids of the Islands and southern provinces," who as soon as "they amended their constitutions," being principally Anglicans, depart for Rhode Island to avoid the "pestilence" of dissent raging in Connecticut. At this time Newport was on its way to becoming a leading summer resort, and Solomon Southwick of the *Mercury* reprinted with obvious solicitude all news items about competitors as far distant as Philadelphia. *Newport Mercury*, June 28, 1773; and the author's "Colonial Newport as a Summer Resort," *Rhode Island Historical Collections XXVI* (January 1933), 1–23.

21. R. G. Albion and Leonidas Dodson, eds., *Philip Vickers Fithian: Journal, 1775–1776* (Princeton, 1934), 112–14.

22. The best contemporary discussion of these springs is in Thomas Jefferson's *Notes on Virginia* (Philadelphia, 1825), 47–50.

23. Samuel Kercheval, *History of the Valley of Virginia* (Woodstock, Va.,

1902), 330; John C. Fitzpatrick, ed., *George Washington: Colonial Traveller* (Indianapolis, 1927), 10, 27.

24. *Virginia Magazine of History and Biography*, XI, 236, 238.
25. *Virginia Magazine*, XIX, 172.
26. Federal Writers' Project, *West Virginia: A Guide to the Mountain State* (New York, 1941), 34, 169; John C. Fitzpatrick, ed., *The Writings of George Washington* (27 vols., Washington, 1931–34), II, 364–65; Fitzpatrick, ed., *George Washington*, 156.
27. In his party were Robert and George William Fairfax, and Peter Kimball of New York, father-in-law of General Thomas Gage. Fitzpatrick, *George Washington*, 246–50; *Writings of Washington*, III, 52.
28. A. T. Volweiler, *George Croghan and the Westward Movement* (Cleveland, 1926), 231; F. N. Mason, ed., *John Norton & Sons* (Richmond, 1937), 189, 263, 275–76.
29. *Fithian Journal*, 123–25.
30. *Fithian Journal*, 126.
31. *Fithian Journal*, 127.
32. Jefferson, *Notes on Virginia*, 48.
33. Fithian visited this spring in January, 1776. The previous month he also rode to Rockbridge Alum Springs in Augusta County, which he learned was "the Resort of many Persons in the Summer." Two other Virginia Springs that came into prominence before the Revolution were the Hot Springs of Augusta County, and the Sweet Springs of Botetourt County. Fithian had evidently been reading Smollett's *The Expedition of Humphry Clinker* (1771), for, in commenting on the taste of the waters at Harrowgate, Matthew Bramble muttered, "Others compare it to the scourings of a foul gun." (*World Classics* edition, Oxford, 1928, 138). *Fithian Journal*, 145–46, 162–63; Jefferson, *Notes on Virginia*, 48–50.
34. The lottery was advertised as early as 1763. Rind's *Virginia Gazette*, May 4, 1763; July 21, 1768; *Fithian Journal*, 162–63; W. W. Hening, ed., *The Statutes at Large . . . of Virginia* (13 vols., Philadelphia, 1819–23), VIII, 16–17, 152, 546–47, 548; IX, 247–48.
35. William Darlington, *Memorials of John Bartram and Humphry Marshall* (Philadelphia, 1849), 338; Carl and Jessica Bridenbaugh, *Rebels and Gentlemen: Philadelphia in the Age of Franklin* (New York, 1942), 315–16.
36. Now called Chester Springs, and situated between Phoenixville and Downingtown.
37. At this time another road to the Springs was under construction. *Bulletin of the Chester County Historical Society* (1916), 9, 10, 37.
38. *Pennsylvania Gazette*, May 19, Nov. 17, 1763; June 7, 1764; July 23, 1767.
39. *Pennsylvania Chronicle*, March 5, 1770; *Pa. Gazette*, March 14, 1771.
40. *Pa. Gazette*, Feb. 9, 24, 1774.
41. *Bull. Chester Cty. Hist. Soc.*, 10–11, 20, 50.

42. *Pennsylvania Magazine of History and Biography*, XLVII, 219; W. W. H. Davis, *History of Bucks County, Pennsylvania* (Doylestown, 1876), 133; William Bache, *Historical Sketches of Bristol Borough* (Bristol, 1853), 10.

43. My italics. *Early Proceedings of the American Philosophical Society* (Philadelphia, 1884), 17, 44; *The Transactions of the American Philosophical Society. Published in the American Magazine during 1769.* (Facsimile, Philadelphia, 1969), 70–76.

44. *Transactions of the American Philosophical Society* (2nd edn., rev., Philadelphia, 1789), I, cases Nos. 1, 5, and 10 are especially interesting.

45. Here was a signal example of Franklin's concept of "philosophy" put to use. The American Philosophical Society not only indulged in cloistered science, but served the public well in advertising Dr. De Normandie's studies, because the indiscriminate use of mineral waters by persons of nearly every colony needed checking. *Trans. Amer. Phil. Soc.*, I, 377–78; *Pennsylvania Journal*, Oct. 6, 1768 (Postscript).

46. W. S. Perry, ed., *Historical Collections of the American Colonial Church: Pennsylvania* (Hartford, 1871), II, 460.

47. These advertisements stressed Dr. De Normandie's account of the Bristol Springs, *Pa. Chronicle*, Apr. 17, 1769; May 21, 1770; *Pa. Gazette*, Apr. 30, 1769; June 11, 1772.

48. Bottled water from Bristol Springs sold throughout Pennsylvania. Benjamin Rush, *Experiments and Observations on the Mineral Waters of Philadelphia, Abington and Bristol* (Philadelphia, 1773), 7–10; for the quotation, 28–30. Rush read this paper before the American Philosophical Society on July 18, 1773. It was referred to the Committee on Natural History.

49. Rush, *Experiments and Observations*, 29.

50. *Pennsylvania Magazine*, XXI, 479; *Pennsylvania Packet*, April 25, 1774.

51. *Pa. Gazette, Pa. Journal*, Aug. 20, 1761.

52. Bath-Town was situated on the north side of Cohocsink Creek, near the Germantown Road and Willow Street.

53. There is a season ticket to the baths in the Hist. Soc. of Pa. *Pa. Journal*, Aug. 22, 1765.

54. *Pa. Gazette*, Aug. 22, 1765; May 15, June 13, 1766; *Pa. Evening Post*, Apr. 1, 1775; *Pa. Packet*, Apr. 3, 1775; *Pa. Magazine*, IV, 179.

55. *Pa. Gazette*, May 9, 1773; *Pa. Chronicle*, May 24, 1773; *Pa. Packet*, May 24, July 13, 1773; *Pa. Magazine*, V, 191.

56. Rush, *Experiments and Observations*, 25–28, 30.

57. Quoted by Cecil K. Drinker, *Not So Long Ago* (New York, 1937), 25.

58. *Pa. Gazette*, May 17, 1767; Oct. 24, 1771.

59. In the writer's youth this resort was called Mineral Springs Hotel; it stood opposite Willow Grove Park. *Pa. Chronicle*, Apr. 11, 1768; *Pa. Gazette*, June 11, 1772; June 22, Aug. 10, 1774; *Pa. Packet*, June 27, 1774.

60. Elizabeth Drinker, Diary (Typescript of Ms in Hist. Soc. of Pa., generously loaned to me by Dr. Cecil K. Drinker of Boston, who used it most felicitously in his *Not So Long Ago*), Sept. 15–18, 1769.
61. Eleazer Wheelock, President of Dartmouth College, and a party, recuperated at Lebanon Springs, N.Y., in 1771. In 1776 Bostonians heard of a "most valuable Spring," to which "great Numbers of People . . . resort . . . daily," at Kinderhook, N.Y. L. B. Richardson, *History of Dartmouth College* (2 vols., Hanover, N.H., 1932), I, 96, 100; *Bos. Post-Boy*, July 21, 1766. For salt-water baths at Manhattan, see *N.Y. Mercury*, Aug. 1, 1768; Nov. 1, 1773; and for Long Island beaches in 1754, Arthur Pound, *Johnson of the Mohawks* (New York, 1930), 85.
62. E. Drinker, Diary, June–August, 1770.
63. E. Drinker, Diary, June–August, 1770; June 26–July 29, 1771.
64. E. Drinker, Diary, June 26–August 11, 1771.
65. E. Drinker, Diary, Aug. 23, 1771; July 1, 1799; C. K. Drinker, *Not So Long Ago*, 29.
66. E. Drinker, Diary, June 9, 27, 1772; May 4, June 1, July 7, 10, 18–20, 1773.
67. E. Drinker, Diary, Oct. 23–29, 1776; Leonard Lundin, *Cockpit of the Revolution* (Princeton, 1940), 42–43, for Stansbury's verses; *Pa. Magazine*, XV, 246, for those of Elizabeth Drinker.
68. Space does not permit discussion of the attempt in 1774 to promote a "public Sea Bath" at Lewes, Delaware; of the Perth Amboy enclosed salt-water baths (1772–73); of the mineral springs of Up-Country South Carolina (1771) which local zealots hoped would prevent Charlestonians from spending expensive summers in the north; of the "Bath Sulphur Springs" in the parish of St. Thomas, Jamaica (1750), which attained 123° F., and proved efficacious for the dry bellyache; or of the "strained sea-water" bath at St. John's Antigua, with its bottom of "polished marble" (1764). *Pa. Gazette*, April 30, 1772; June 11, 1773; Feb. 23, 1774; *N.Y. Gazette, or the Weekly Post-Boy*, June 22, 1772; D. D. Wallace, *The History of South Carolina* (3 vols., Chicago, 1934), II, 103; Bryan Edwards, *The History Civil and Commercial of the British West Indies* (2 vols., London, 1794), I, 193; C. M. and E. W. Andrews, eds., *Journal of a Lady of Quality* (New Haven, 1927), 111.
69. John Rouelle, *A Complete Treatise on the Mineral Waters of Virginia* (Philadelphia, 1792), 64.
70. *Pa. Magazine*, LX, 485.